Environmental Management for Aquaculture

CHAPMAN & HALL AQUACULTURE SERIES

Series Editor: Michael G. Poxton, Aquaculture Associates International, The Firs, Glasshouse Hill, Codnor, Derbyshire, DE5 9QT.

During the last decade aquaculture production has greatly increased. Over 26% of all shrimp and 20% of salmon consumed are now farm grown, both up from only 1% 10 years ago. Indeed, this dynamic world-wide industry is now growing at nearly 10% annually.

Although extensive culture in Asia is still extremely important in terms of overall production it is the culture of expensive commodities such as salmon and shrimp in intensive and semi-intensive systems that has aroused most recent interest. Cultures of this sort have focused attention on the need for new diets, new equipment, new management practices and environmental impacts. A tremendous impetus has been given to the industry by a flood of research papers in the last two decades, yet it remains the case that there are still relatively few universities offering well designed courses in aquaculture and relatively few aquaculture textbooks, manuals and comprehensive but succinct and affordable treatises on which such courses can draw support.

The purpose of this aquaculture book series is to present timely volumes reviewing important areas of aquaculture and to encourage authors to synthesize the bulging literature and distil the information to provide common-sense, practical advice and accessibility. The level of the series will be suitable for undergraduate students in biology, bioengineering, biotechnology, aquatic sciences, physiology and zoology and the applied nature of the topics covered will appeal to a wide range of fisheries scientists, consultants, fish farmers and managers. By being both topical and up to date they should also be of interest to researchers who require to keep abreast of developments in what is after all the ultimate multidisciplinary subject.

Available titles
1 Fish Nutrition in Aquaculture
Sena S. De Silva, Deakin University, Victoria, Australia and **Trevor A. Anderson**, James Cook University of North Queensland, Australia.
December 1994; 340pp. Hb: 0 412 55030 X.
2 Environmental Management for Aquaculture
Alex Midlen, Consultant in Aquaculture, East Riding of Yorkshire, UK and **Theresa A. Redding**, University of Hull International Fisheries Institute, Kingston-upon-Hull, UK.
May 1998; 240pp. Hb: 0 412 59500 1.

Forthcoming titles
Water Quality for Fish Culture
Michael Poxton, President, European Aquaculture Society.
1998: c.352pp Hb: 0 412 44860 2.
Applied Fish Pharmacology
Keith Treves-Brown, MAFF Veterinary Medicines Directorate, Surrey, UK.
1998: c.350pp Hb: 0 412 62180 0.
Fish Larval Nutrition
Edited by **Johan Verreth**, University of Wageningen, Netherlands.
1999: c.265pp Hb: 0 412 62110 X.

Publisher: Nigel J. Balmforth, Chapman & Hall, 2–6 Boundary Row, London SE1 8HN

Environmental Management for Aquaculture

Alex Midlen
Consultant in Aquaculture
East Riding of Yorkshire, UK

and

Theresa A. Redding
University of Hull International Fisheries Institute
Kingston-upon-Hull, UK

CHAPMAN & HALL
London · Weinheim · New York · Philadelphia

Published by Chapman & Hall, an imprint of Thomson Science, 2–6 Boundary Row, London SE1 8HN, UK

Thomson Science, 2–6 Boundary Row, London SE1 8HN, UK

Thomson Science, 115 Fifth Avenue, New York, NY 10003, USA

Thomson Science, Suite 750, 400 Market Street, Philadelphia, PA 19106, USA

Thomson Science, Pappelallee 3, 69469 Weinheim, Germany

First edition 1998

Thomson Science is a division of International Thomson Publishing

Typeset in 10/12pt Palatino by Florencetype Ltd, Stoodleigh, Devon
Printed in Great Britain by St Edmundsbury Press, Bury St Edmunds, Suffolk

ISBN 0 412 59500 1

A catalogue record for this book is available from the British Library

Library of Congress Catalog Card Number: 98–70278

Contents

Series editor's preface

Environmental Management for Aquaculture focuses attention on the methods of production and the environmental consequences of the production process in order that negative environmental impacts can be minimized. The basic theory of 'systems ecology' is introduced and applied to culture production systems in order that different production processes may be compared against a common framework, principles and reference points, the basis of which are energy transformation and nutrient cycling.

The authors avoid confusing the reader with too much detail on pollution control technology or fish nutrition as suitable texts already exist. The unique aspect of *Environmental Management for Aquaculture* is its discussion of all these facets of aquaculture within the framework of environmental management. Similarly, the authors have not presented an exhaustive catalogue of observed impacts, preferring instead a series of illustrative examples and case studies which together provide an overview of environmental effects. Methods of limiting the generation of polluting substances are explored while pollutant removal technologies, especially biological water treatment systems, are also considered.

Alternative technologies for more sustainable aquaculture production systems are introduced and strategic approaches to environmental management are developed taking account of the rapidly changing regulation of environmental impacts that are being applied in various countries.

The authors' easy style makes these often complex issues readily available and as such this most recent addition to Chapman & Hall's Aquaculture Series will be appreciated by a wide audience.

Michael G. Poxton
Series Editor

Preface

The impact of aquaculture on the environment can be manifest in a wide variety of ways. The most widely studied are the polluting effects of waste products from the production process, but of no less importance are other forms of impact. These include visual amenity, conflict with other uses of renewable natural resources (mainly leisure, fishing, and mangrove, wetland and wildlife conservation), introduction of exotic species and genetic pollution of wild stocks. This wide range of issues covering social, economic and environmental impacts is beyond the scope of this book, which will instead concentrate on the alleviation of the polluting effects of aquaculture through environmental management. The authors wish to stress, however, that these other factors must also be given full weight in any environmental impact assessment and environmental management system. The approach illustrated in this book for the application of environmental management principles by reference to pollution control is equally valid with regard to other impacts. A useful introduction to the nature of these other impacts is given in Pillay (1992).

This book covers a broad range of subjects, from systems ecology to sedimentation theory to environmental policy. The treatment of each topic is consequently of an introductory nature, although it is hoped that enough detail is given to assist in the development of strategies for environmental management of aquaculture systems and of regional aquaculture development. Where there is an obvious need for more information, the reader is directed to appropriate sources. Case studies are used frequently within the text, in the belief that they are a valuable way of imparting an intuitive understanding of the subject, as well as providing a vehicle for factual information.

The authors would like to pay tribute to all those whose hard work we have drawn upon in the writing of this book, and to those who have contributed to its production.

CHAPTER 1

Aquaculture systems

1.1 INTRODUCTION

Aquaculture – the farming of aquatic animals and plants – is an important industry worldwide. The production systems used differ widely depending on the species being cultured and on the geographical location and socioeconomic context. However, all forms of aquaculture have one thing in common, i.e. the utilization of natural resources and the manipulation of biological systems. Any such exploitation must conform to the fundamental 'rules' governing biological processes: the two laws of thermodynamics, which state that:

- energy cannot be created or destroyed within a system but may be changed into different forms (or, 'what goes in must go out');
- in a system where energy is transformed (e.g. from food to flesh) there is a degradation or loss of energy in the form of heat (or, in other words, no biological system is 100% efficient).

These laws refer to the patterns of material recycling and energy flow which are common to all cycles of life (Odum, 1993). Thus most aquaculture systems rely on a supply of nutrients and energy to maintain their productivity, and in turn produce waste nutrients and waste energy. These wastes, which are potential pollutants, do not arise solely as a result of human inefficiency in exploiting natural systems: a proportion is an inevitable consequence of those processes that sustain life, as described by the laws of thermodynamics. It is for this reason that environmental management, embodying waste minimization and pollution control systems, is an essential part of the management process. Without it the effects of these wastes are likely to damage the interests of the aquaculture industry. The purpose of this book is to explain why this should be so, and how environmental management can be used to ensure the sustainability of aquaculture.

The extent of pollution from aquaculture systems varies widely depending on the level of technology employed and the management of the system. This introductory chapter reviews aquaculture systems and their potential to pollute natural waters. (For those who are unfamiliar with the technology of aquaculture, the following section will give sufficient information to enable an understanding of the rest of the book.)

1.2 A CLASSIFICATION OF AQUACULTURE

It has been estimated that in excess of 20 000 different species of fish have been described. Of these, only about 100 are cultured for commercial purposes (Shepherd and Bromage, 1988). These figures do not include other aquatic animals and plants that are also farmed. The most important culture techniques are reviewed in later sections. However, whilst the physical nature of the culture system is important with regard to pollution, its management is perhaps of more significance. One key management feature of aquaculture systems is the **intensity** of production.

The term intensity refers to a combination of two variables: the amount of biological material harvested per unit area of the culture system (its productivity) and the degree of manipulation of natural processes (its technological status). All present aquaculture systems can be defined with regard to intensity and placed on a continuum from extensive to super-intensive (Pillay, 1990). The use of the term super-intensive has arisen because the intensity and technological status of aquaculture has progressed in the last 15 years beyond the original definition of intensive.

1.2.1 Extensive coastal aquaculture

Perhaps the best illustration of extensive aquaculture is provided by the traditional systems for farming milkfish (*Chanos chanos*) in Indonesia and the Philippines (Bardach *et al.*, 1972). Milkfish spawn once or twice a year near the coast, and the larvae develop in the coastal and estuarine waters. It is thought that the farming of milkfish may have its origins in the coastal salt pans, which were filled with seawater, then allowed to dry out so that salt could be collected. Now farmers no longer rely on 'accidental' stocking of the milkfish fry in the coastal ponds (tambaks) in Indonesia. Instead there is a well developed supply network for milkfish fry caught in shallow coastal waters.

The fry are stocked into ponds that have been prepared through the addition of fertilizer to enhance the growth of mats of blue-green algae on the pond bottom. This provides an ample source of food for the milkfish fry. The nursery ponds are usually relatively small, and once the milkfish have reached about 5–7 cm in length (which takes about 30 days) they are transferred to larger ponds for on-growing to market size (Figure 1.1). These ponds are prepared in a similar way, but are also fertilized with green manures (such as mangrove leaves) to ensure a

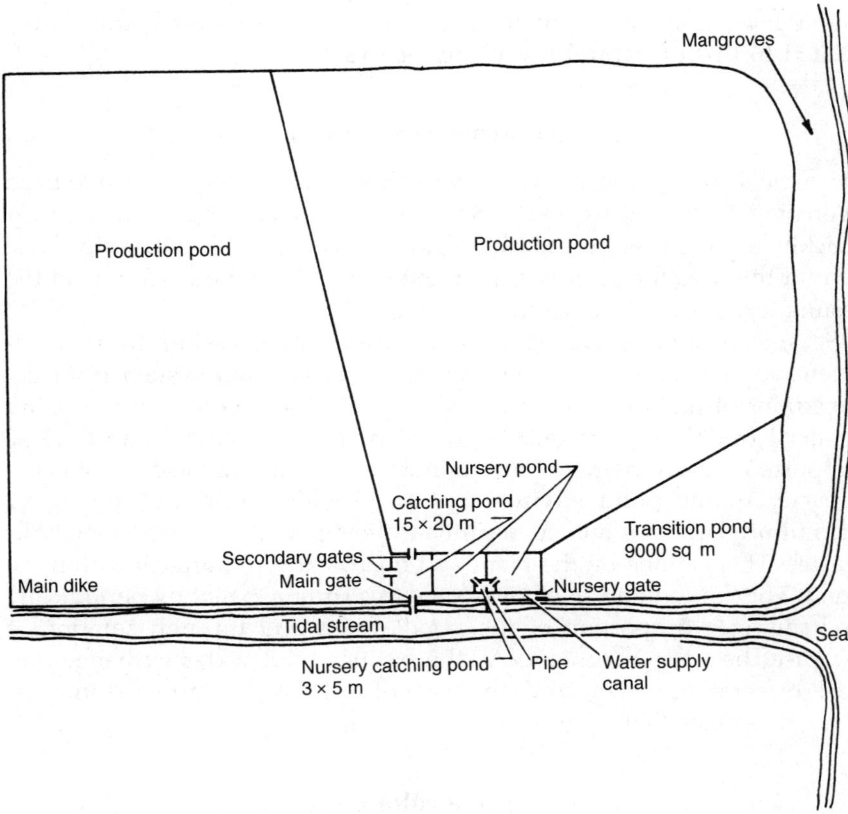

Figure 1.1 Milkfish pond system in the Philippines. (From Bardach *et al.*, 1972.)

luxuriant growth of blue-green algae throughout the on-growing period. Supplementary foods are not usually supplied.

1.2.2 Semi-extensive carp polyculture

The idea of fertilizing fish ponds is thought to have been developed initially by the Chinese. Human and animal manures, and plant wastes, are used to raise the productivity of waters beyond natural levels. Carp fry are produced in hatcheries, then transferred to nursery ponds in which a sufficient supply of live food has been provided by fertilizing the pond beforehand. As the fish grow, the ponds are fertilized regularly to maintain levels of productivity and thus ensure good fish growth. In addition to the live food in the pond, the fish are provided with some supplementary food, such as cereal grains, to improve further the food supply and the productivity of the pond (Pillay, 1990). Despite this higher level of management, the productivity of the system is still closely related to that of natural freshwater ecosystems.

1.2.3 Semi-intensive shrimp farming

Once the technical constraints to the farming of marine shrimp had been overcome in the 1970s and 1980s, the shrimp farming industry very quickly adopted semi-intensive farming systems. The driving force behind this was the rapidly expanding demand for the product and the resultant incentive to maximize output.

Semi-intensive aquaculture is a natural progression from semi-extensive techniques. In order to free the production system from the constraint of natural food production within the ponds, higher quality artificial feeds are provided. A greater biomass of shrimp can thus be supported within the pond environment. The limiting factor now is the capacity of the pond environment to provide sufficient oxygen for the cultured species and to assimilate their potentially toxic metabolic wastes. This is done by the plant and microbial communities within the pond. This limitation can be overcome to a certain extent by periodically exchanging a proportion of the pond water, and through aeration to increase the oxygen content. As the frequency of water exchange and level of aeration is increased, the level of technology employed increasingly resembles that of an intensive system.

1.2.4 Intensive culture systems

In intensive aquaculture systems a continuous flow of water (or frequent water exchange in the case of shrimp) is required to maintain the water quality in the culture system at an acceptable level. High quality artificial feeds are required to provide all the nutritional requirements of the cultured species, and the system must be closely monitored. The supply of food to the fish is now completely under human control, and few ecological factors influence the biomass of fish that can be supported by the pond, tank or cage. The main limitations are the flow of water available and the tolerance of the cultured species to crowding.

1.2.5 Super-intensive recycle systems

By installing full water-reconditioning systems, the culture system is essentially independent of biological constraints, except those imposed by the biology of the cultured species. The ability to provide artificial environments by reconditioning the water (heating or cooling it, providing artificial lighting regimes and artificial foods) enables the production of almost any species at any location. The only constraint then becomes the economic viability of the enterprise. Such systems are expensive to construct and operate, and are suited to the culture of species that are high in value or that are tolerant of high stocking densities. This reflects the fact that the income per unit of production plant must be maximized for these systems to be profitable.

1.3 PHYSICAL NATURE OF AQUACULTURE SYSTEMS

The level of intensity is one of the basic characteristics of any aquaculture system. Another is the physical nature of the system. Aquaculture stocks are usually contained during their culture period. This is to prevent escape, predation or theft, and also to ensure that the organisms remain in a suitable environment. The need for this is obvious in the case of finfish which are highly mobile, but it is equally important for sessile animals and plants which do not move. The provision of substrates for attachment in a suitable environment is necessary for the successful culture of sessile organisms. In the following paragraphs the most widely encountered culture systems are dealt with in turn.

1.3.1 Earth ponds

Earth ponds are the most widely used system for aquaculture – they are relatively cheap to construct and maintain, and they are a 'low-tech' and therefore widely accessible solution to the problem of confinement of finfish, crustacea or single-cell algae.

Earth ponds can be categorized with respect to the water requirements of the culture system. Species such as trout require a continuous flow of water through the pond in order to renew the supply of oxygen and to remove the waste products of metabolism. Other species are adapted to different conditions. Carp, for example, are usually farmed in static-water ponds that are fertilized to increase the production of natural food organisms.

Fish farms with earth ponds or constructed tanks usually have a single inlet point at which water is drawn from a river or canal and distributed around the farm. Water leaving the ponds, whether continuously or periodically, usually enters a common collection channel and flows through a sedimentation pond before being discharged from the site (Figure 1.2).

1.3.2 Tanks

Earth ponds do suffer from some disadvantages for culturing those species which need continuous water flows. The most important problem is that the flow of water through the pond may not circulate to all areas, leading to partial stagnation. This effectively restricts the intensity of production to below the maximum level. Equally important is that earth

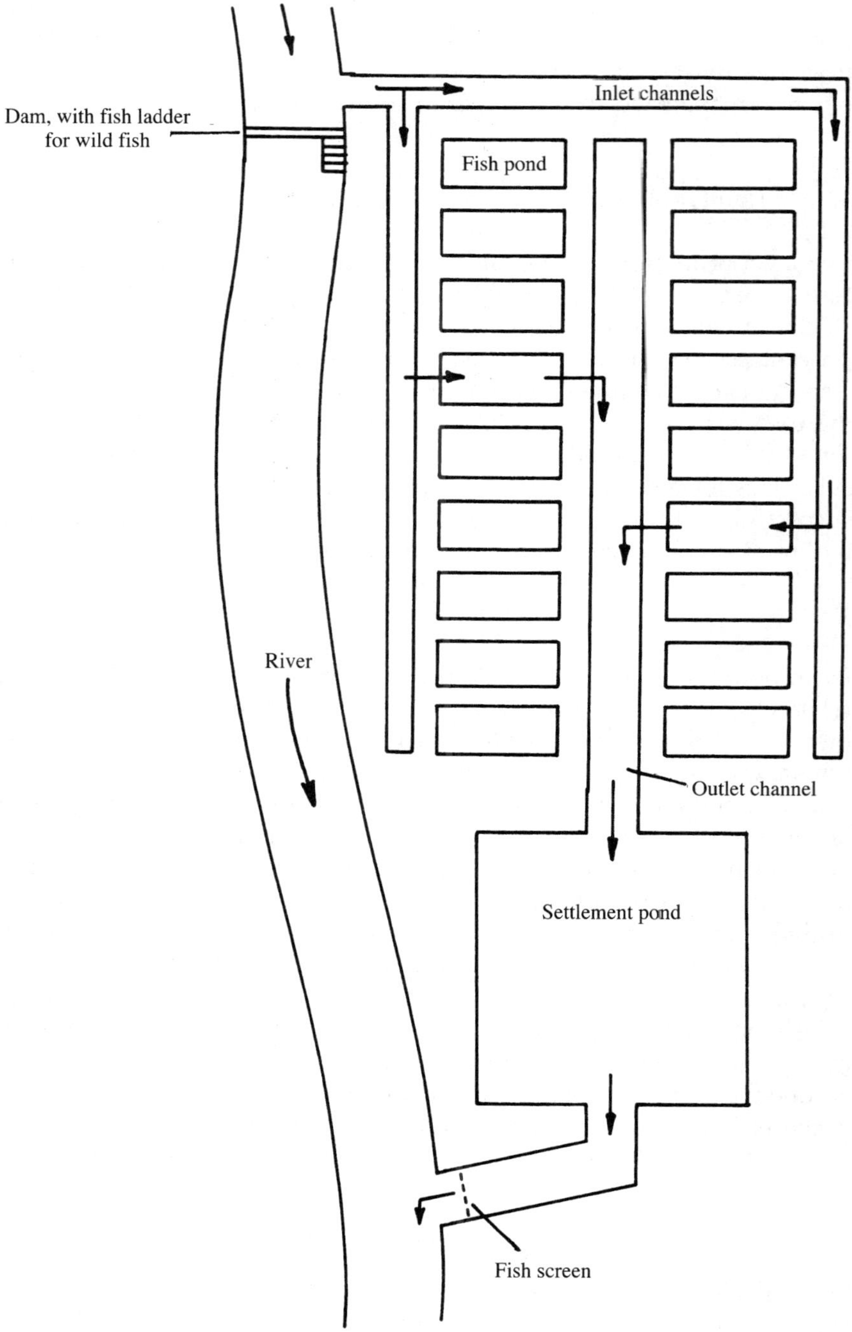

Figure 1.2 Early Danish layout for earth ponds, with a settlement pond for solids removal. (Modified from Sedgewick, 1990.)

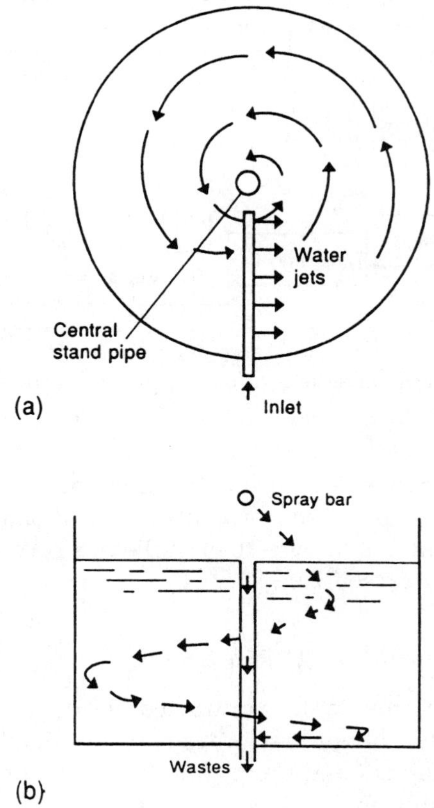

Figure 1.3 Water circulation patterns in a circular tank: **(a)** plan view; **(b)** elevated view. (From Lawson, 1995.)

ponds with flowing water are prone to bank erosion and siltation, which may result in poor water quality and increased disease risk.

The development and use of tanks may have overcome some of these problems. Most designs have better hydraulic characteristics than earth ponds, thus enhancing oxygen supply and the removal of waste metabolites. Self-cleaning tanks generally conform to one of two designs: the circular tank and the raceway.

(a) Circular tanks

In the circular tank a spiral water flow is generated by the force of the inlet water and the slope of the bottom (Lawson, 1995). This ensures that there are no stagnant pockets of water in the tank, and that any solids are swept along in the fast current to the outlet in the centre of the tank floor (Figure 1.3). Circular tanks may be prefabricated in sections

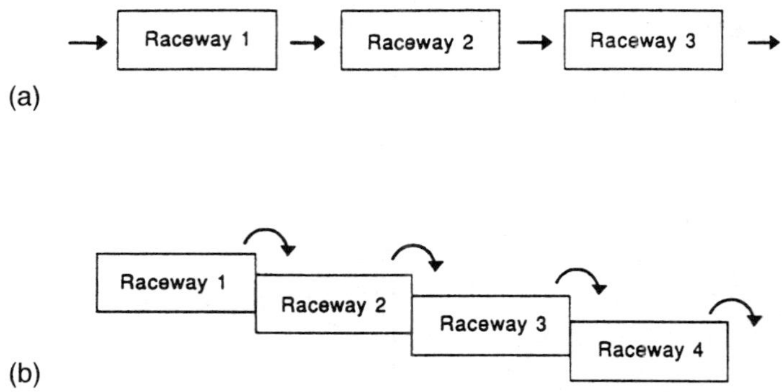

Figure 1.4 Raceway units in series: **(a)** on flat ground; **(b)** on sloping ground. (From Lawson, 1995.)

from fibreglass and assembled on site; they may be of plastic or rubber-lined wood or corrugated steel, or constructed *in situ* from concrete. Sizes vary from about 1 m to over 10 m in diameter. Water depth varies but does not usually exceed 2 m.

(b) Raceways

Raceways are long narrow tanks constructed *in situ*, usually of concrete. The flow rate of water through raceways is very fast, allowing high fish densities to be maintained and preventing the collection of solid wastes within the culture unit (Figure 1.4).

1.3.3 Floating net cages

The development of floating net cages has allowed the exploitation of lakes, rivers, canals, estuaries and the sea for aquaculture. They commonly comprise a floating collar – square, octagonal or circular in shape – from which hangs a net bag in which the fish are kept (Figure 1.5) (Beveridge, 1987). Water exchange is required, in the same way as for ponds and tanks, to maintain a healthy environment for the fish. This is ensured by siting the cage where there is sufficient water current to flush the net bags at an appropriate rate. Nets are prone to fouling (the growth of invertebrate animals and aquatic plants on the net material), which blocks the meshes and reduces the rate of water exchange. Consequently, they must be periodically changed and cleaned.

1.3.4 Longline and raft culture

Some bivalve molluscs and seaweeds grow attached to a hard substrate, such as rocks, in the wild. There is not the same problem of containment

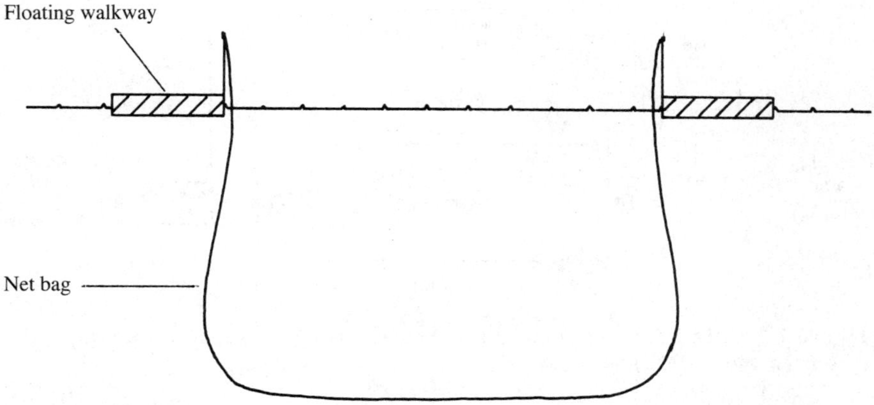

Figure 1.5 A floating net cage.

as there is with fish and crustacea, but the provision of artificial substrates for the early life stages of these organisms to settle on and attach to affords the opportunity to move them to the most favourable growing environments.

In the traditional French system, mussel spat are collected on ropes wound around vertical poles set in the intertidal zone. More recently, floating longline or raft systems have been developed for bivalve culture (Figure 1.6). These have allowed the utilization of many more sites.

Scallop culture has also been developed using floating systems. Oysters, on the other hand, are more usually cultured on the sea bed in the intertidal zone. The methods used vary from traditional enclosures, in which oysters are set on the sediment surface, to vertically stacked tray systems.

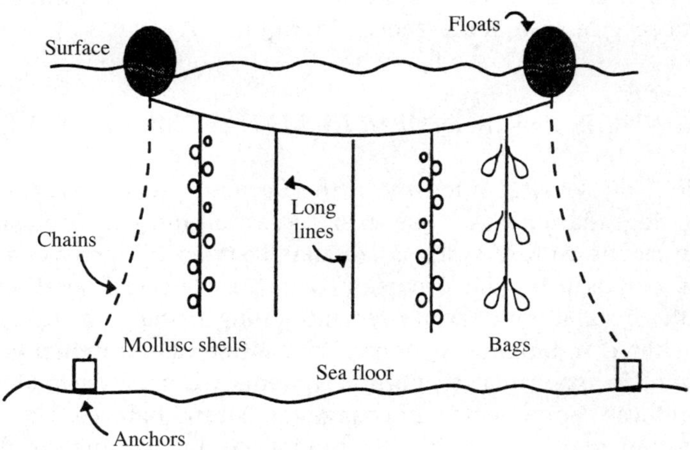

Figure 1.6 Longline culture of shellfish. (From Lawson, 1995.)

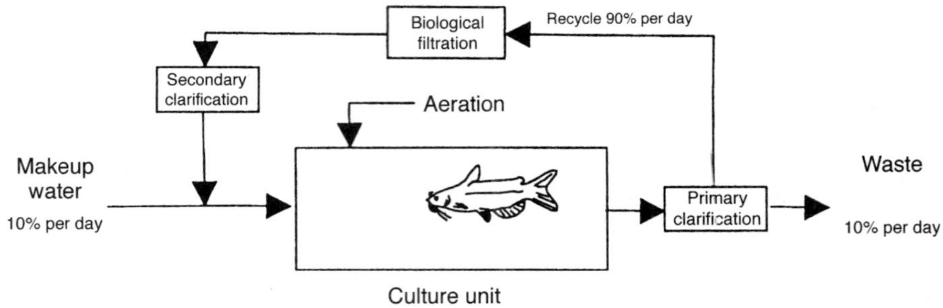

Figure 1.7 A recirculating aquaculture system using 90% recycle per day. (From Lawson, 1995.)

Seaweeds have similar requirements to mussels with regard to substrates and are cultured on suspended nets or ropes.

1.3.5 Recycle systems

The ultimate expression of intensive fish farming is the recycle system, in which an almost totally artificial environment is provided for the fish. As much as 95–99% of the water flow is recycled, being filtered and reconditioned before returning to the stock tanks, and temperature and light regimes are artificially controlled (Lawson, 1995).

Recycle systems have the important benefit of independence from voluminous water supplies, in contrast to most intensive culture systems, and of allowing full environmental control of the culture system. Thus tropical fish can be cultured in colder climates, and light regimes can be controlled to allow manipulation of reproductive cycles. This is important in ensuring year-round supplies of fish. The components of a typical recycle system are shown in Figure 1.7.

1.4 WHY IS ENVIRONMENTAL MANAGEMENT NEEDED?

How does the level of intensity of the farming process cause environmental degradation? As one progresses along the intensification continuum, the culture systems are characterized by increasing manipulations of Nature – the provision of artificial foods, aeration of the water, the installation of water-reconditioning systems. Natural ecosystems achieve a balance between the different components of the ecosystem. In manipulating natural systems we alter the relationships between these components and change the natural balance. The majority of environmental degradation attributable to aquaculture results from imbalances within the underlying ecosystem processes of energy flow and nutrient cycling. The use of chemicals in the farming process may also be significant. Mostly used to combat disease, these chemicals are often highly toxic to wildlife and may be the direct cause of environmental degradation. These effects can be minimized by the process of environmental management, which focuses attention on the environmental consequences of the production process.

1.5 SYSTEMS ECOLOGY AND ENVIRONMENTAL MANAGEMENT

Different processes have different attributes with regard to what goes into the process and what comes out of it. This makes comparison between processes difficult. What is needed is a framework based on common principles and common reference points, so that each process can be taken back to basics, or 'first principles'. There is such a framework available, based on the key processes of natural ecosystems – energy transformation and nutrient recycling. This is the systems ecological approach. In the following sections the basic theory of systems ecology will be introduced, followed by its application to aquaculture systems. These concepts will be referred to in subsequent chapters.

1.5.1 Systems ecology theory

The word 'system' can be used to describe both natural and industrial processes. The *Oxford English Dictionary* defines a system as 'a group, set or aggregate of things, natural or artificial, forming a connected or complex whole'.

Some examples are: a thermostatically controlled central heating system; the human central nervous system; a city; a fish farm; an ecosystem (a lake, the oceans, a desert, a tropical rainforest) (Figure 1.8).

Systems characteristically have (Odum, 1993):

- inputs from outside the system;
- a processing function within it;
- outputs to the external environment.

There are also feedback mechanisms which control the processing function, such that outputs from the system modify the surrounding environment so as to modify future inputs. The purpose of the feedback mechanisms is to maintain stability within the processing function. If one considers a real example these characteristics should become much clearer.

A thermostatically controlled central heating system is the classic example of an engineering system. It displays all the essential characteristics of a system, and is easily understood. The purpose of a central heating system is to keep a house at a certain temperature, usually higher

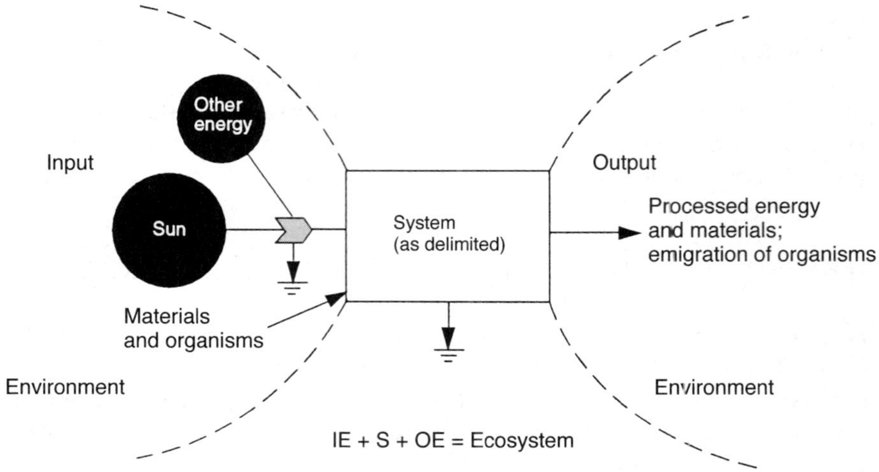

Figure 1.8 Model of an ecosystem as an open, thermodynamic non-equilibrium system, with emphasis on the external environment, which must be considered an integral part of the ecosystem concept. (Redrawn from Odum, 1993.)

than the temperature outside. Thus, the system comprises a boiler which converts incoming energy, in the form of oil, gas or coal, into heat. The heat is transferred to the rooms by convection and radiation from the surface of radiators, and this raises the air temperature. If this system were unregulated, the temperature would just continue to rise. To keep the air temperature at a comfortable level, therefore, a thermostat is incorporated into the system. If the room temperature rises above a specified level (a result of the output of heat from the system), the thermostat causes the system to shut down, i.e. to reduce its heat output. If the air temperature then falls below the desired temperature, the thermostat causes the system to increase its heat output. Thus the thermostat regulates the system so as to maintain a constant temperature in the room. This is a self-regulating, homeostatic system. In warmer climates, an air-conditioning system works on the same principles, but the processing function is a cooling process.

Feedback of the type described above is called **negative feedback**; that is, when a rise in output triggers a restriction on the flow of inputs to the system (fuel, in this example). Homeostatic systems often have a limited range of conditions in which they can function. Outside these limits the self-regulation breaks down. This can lead to **positive feedback**, where an increase in output can trigger an increase in inputs.

The mammalian temperature control mechanism works quite like the central heating system. The chemical reactions in the body generate heat as a by-product. Food (fuel) is the input, and heat is the output from the system. When the human body becomes hot, sweat is produced. This is negative feedback, triggering a cooling mechanism. When the body is cold, shivering generates heat. By these mechanisms the body is able to maintain a constant temperature. However, if the body overheats, the negative feedback system breaks down. The metabolic rate is elevated to such a level by the external temperature that the body's cooling mechanisms cannot cope with the extra heat produced. The body then becomes warmer, the metabolic rate and consequent heat production rises further, which increases the body temperature to a level where death eventually occurs from heat stroke. Thus, if the external environment exceeds certain limits, the system will break down.

The systems approach can be applied to the functioning of natural systems (ecosystems) and to semi-natural and artificial systems (farming and industrial systems) (Odum, 1993).

An ecosystem is one in which biological and physical processes work together to maintain a complex web of life (Tivy and O'Hare, 1981). An ecosystem can be any size – a puddle or the Atlantic ocean – but the tendency is to look at ecosystems that have clearly defined physical and biological characteristics. Thus, it is usual to consider a forest or a heather moor or a desert as separate ecosystems. In the aquatic environment, too, there are a variety of separate ecosystems such as rivers, lakes, the oceans, marshes and mangrove swamps. As with any system these ecosystems have characteristic inputs, processing functions, outputs and feedback mechanisms. These are much more complex than the examples given to illustrate these characteristics; nevertheless, certain key processes in each ecosystem type, their corresponding inputs and outputs, and how the feedback mechanisms work can be defined. This understanding can then be used to evaluate and minimize the impact of human activities on the environment. This will be considered in more detail later, but first inputs and processing within the ecosystem are described in the following sections.

There are two categories of input to ecosystems: energy and minerals. These two categories differ in one vital respect: energy cannot be recycled within the system, but minerals can (Odum, 1993). Regarding minerals, a distinction can be drawn between those that are incorporated into organic material, which are referred to as bioelements, and those that are not, which shall be referred to as minerals.

When energy enters an ecosystem it is processed, i.e. converted into a different form. This conversion process cannot be 100% efficient. During the conversion process in a biological system some of the energy is lost as heat. Figure 1.9 illustrates this gradual degradation of energy as it progresses through different biological components of an ecosystem. Once it has been degraded the heat energy cannot be recovered. Thus the Earth depends on a constant supply of energy from the sun to enable all systems, such as the weather and the oceans, to continue functioning since all are ultimately powered by the sun's energy.

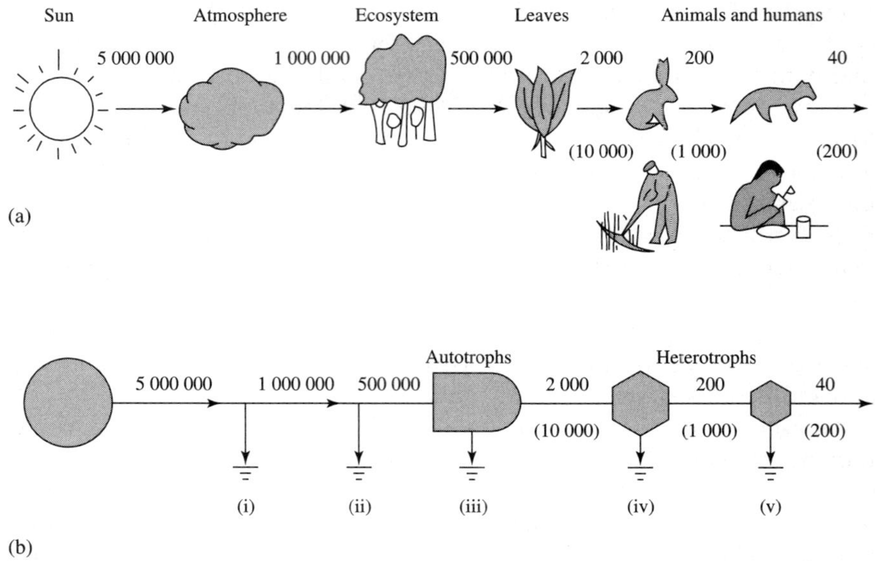

Figure 1.9 Solar energy flow (kcal m^{-2} per year) through the biological foodchain: **(a)** pictorial diagram; **(b)** more formal energy flow model. Figures in parentheses show levels that may be reached in subsidized ecosystems when solar energy flow is enhanced by other types of energy such as fuels. (Redrawn from Odum, 1993.)

Minerals, by contrast, do not degrade. They may be transformed in some way, but they remain available to the different processes within the ecosystem, unless transported out of the system. Thus, when a plant dies its leaves and stems are decomposed by bacteria and other organisms. The bioelements originally contained in the plant are eventually released into the soil, where they are again available to be incorporated into new plant tissue. In other words, they are recycled.

Different elements have different cycles, depending on their physical and chemical characteristics (Figures 1.10 and 1.11). Nitrogen and phosphorus are very important bioelements because it is these that are most often in short supply in the environment in a 'free' form, i.e. not already incorporated into living material. Plants utilize bioelements when they occur in particular ratios to each other, which is a result of the way in which they are incorporated into organic material. If one bioelement is in short supply the 'building work' of organic growth can be completed only at the rate at which that one is supplied, no matter how much of all the other bioelements are available. Such bioelements are known as limiting nutrients. In seawater nitrogen is normally the limiting nutrient, and in freshwater it is usually phosphorus, but both are very important in any environment because of their vital roles in plant and animal life.

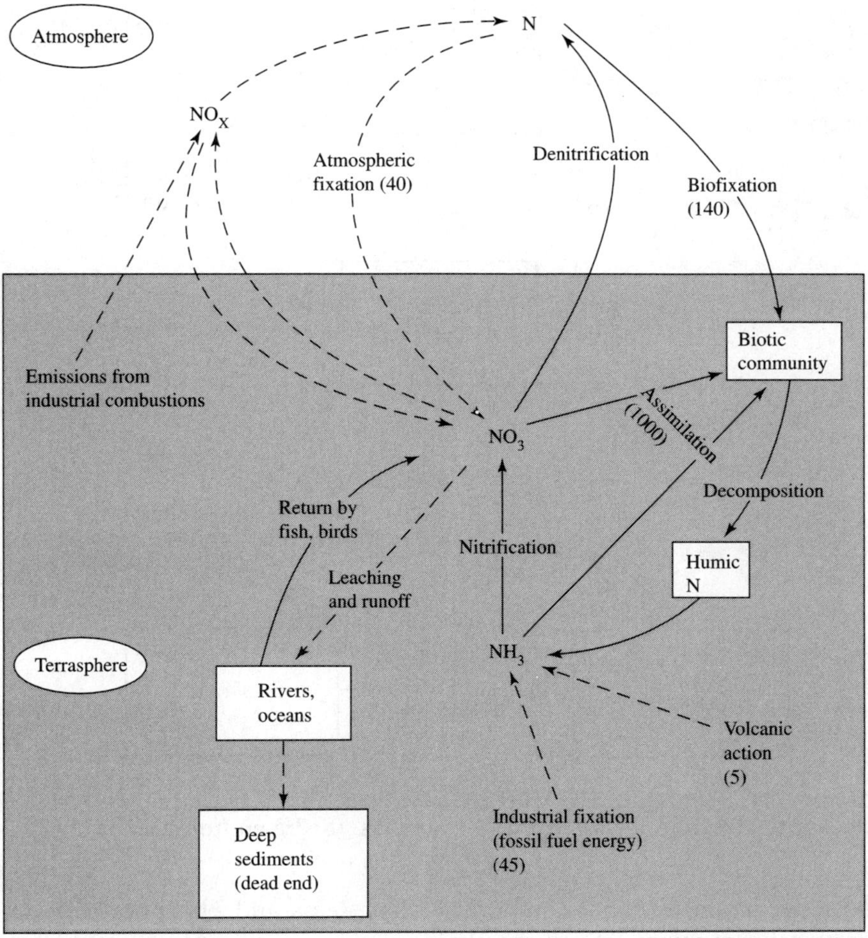

Figure 1.10 The nitrogen cycle. Nitrogen cycles in its several major forms between the large storage pool in the atmosphere and the smaller, but more active, pools in the soils and water of the earth itself. Solid arrows indicate the flows and exchanges mediated and controlled by organisms (especially microorganisms). Dashed lines show the flows that result chiefly from physical forces or are the result of human activities. Nitrogen oxides (NO_x) in the atmosphere can be in several forms, such as NO_2 or N_2O. These are the forms that contribute to acid rain, smog and other forms of air pollution. At present it is not possible to put numbers on all the flows at the global level. The numbers in parentheses are there merely to show the relative importance of some exchanges – for example, biofixation by nitrogen-fixing microorganisms is more important than fixation by lightning or industrial fixation during the production of fertilizers. (Redrawn from Odum, 1993.)

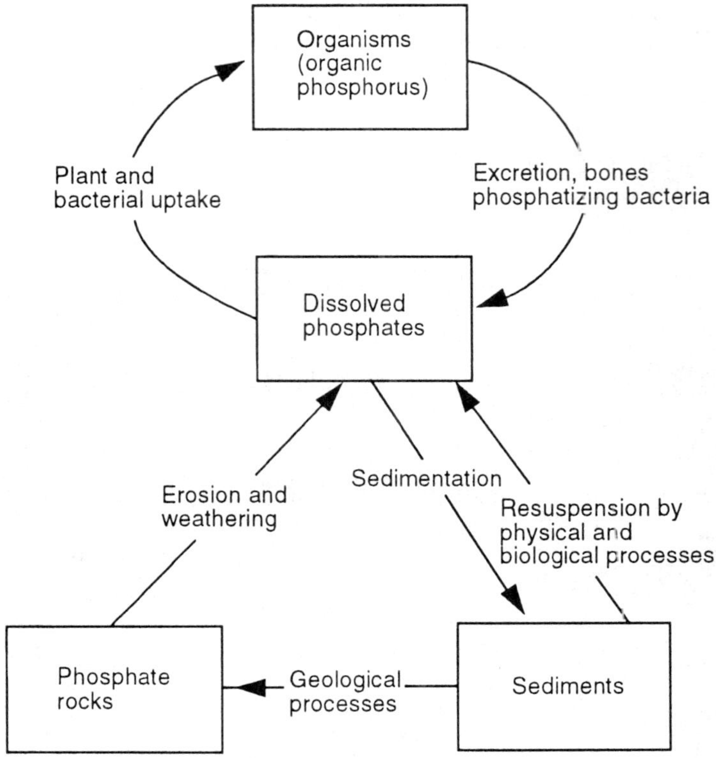

Figure 1.11 Simplified diagram of the phosphorus cycle, including slow fluxing to and from the sedimentary reservoirs. (Redrawn from Odum, 1993.)

Nitrogen is an essential component of proteins and phosphorus is an essential component of cell membranes.

A river has a variety of energy sources. Most obvious is sunlight energy, which is captured by plants such as algae and larger aquatic plants. Another source of energy is dead organic material – leaves, twigs, dead animals – which enters from the land and floats downstream. The minerals entering the system also come from a variety of sources. All the organic material coming from outside the system (in the water flowing downstream, by washing from the land, or to a certain extent in rainfall) contains a variety of minerals in dissolved or particulate form. These inputs, particularly that of energy, are vital to the functioning of the ecosystem. They support the lives of the animals and plants that inhabit it. As with the central heating system, there must be feedback mechanisms to maintain stability in the ecosystem. Once the ecosystem is seriously disturbed, overloaded with inputs for example, the feedback mechanisms cannot cope. This may have serious environmental consequences.

In the case of aquatic environments, overloading the system with nutrients and energy (by discharging nutrient-rich wastewater, for example) causes eutrophication. This is a long-term rise in the level of nutrients throughout the ecosystem and a resultant shift in its ecological balance. This is a sign of ecosystem stress – that the ecosystem is nearing the limit of its capacity to self-regulate (Tivy and O'Hare, 1981). Once this limit has been breached the stability of the system is often lost quite spectacularly, with massive algal blooms and crashes. Subsequent decomposition of the algal population causes deoxygenation of the environment, and often severe loss of life amongst the higher animals (invertebrates, fish) and larger aquatic plants. In the oceans this can also affect sea bird and seal populations as their food supplies are decimated.

1.5.2 Systems ecology and aquacultural systems

In this section two examples of aquaculture operations are considered from the systems point of view. In doing this the aim is to illustrate how aquaculture can cause instability in ecosystems, and on a more positive note to offer insights as to how one might go about minimizing that impact.

(a) The intensive fish farm as a system

Looking at a fish from a systems point of view: energy, carbon, nitrogen and phosphorus are taken in as food (so are other elements, but these are the most important nutrients and have the greatest impact on the environment) and processed within the body. Oxygen is taken from the water across the gills for respiration. The corresponding outputs are excretory wastes – faeces and urine (which themselves contain C, N and P) – and respiratory wastes (carbon dioxide and ammonia-N). Of the input energy and nutrients, not all are excreted: a significant proportion is retained in the body in the form of tissue growth.

This model, with modification, can be used to represent a fish farming system. This approach was used in a Polish study to investigate the impact of cage farming of trout in a freshwater lake (Penczak *et al.*, 1982). The flow of nutrients and energy through the system is shown in Figure 1.12. Fish feed composition and feeding methods have improved considerably since this study was undertaken, and the level of wastes from the farm is high by today's standards. Nevertheless, it represents a good example of the systems ecology approach in aquaculture research.

The scale of the ecological inefficiency of this production system can be appreciated from such a model. The aspects that make it inefficient are:

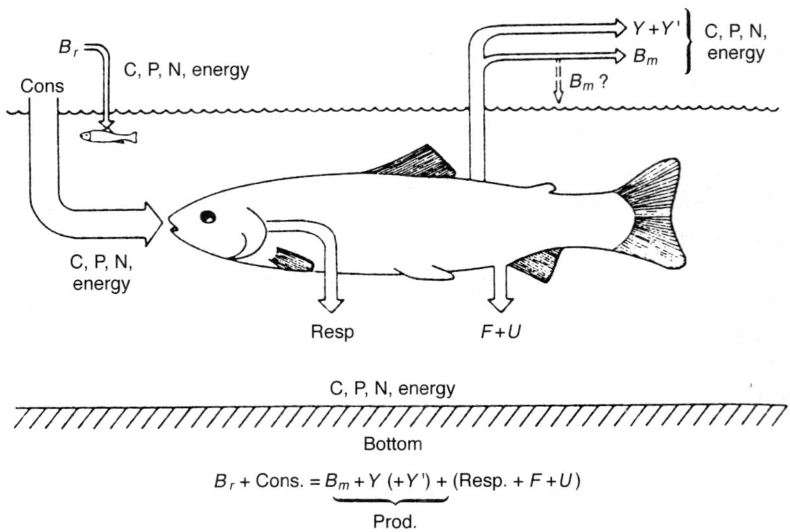

Figure 1.12 Diagram of the flow of energy and bioelements (C, P, N) through the cage aquaculture of rainbow trout in Glebokie Lake, Poland. B_r, recruitment biomass; **Cons.**, total energy consumed; B_m, mortality biomass; ***Y***, energy content of harvested 1+ trout; **Resp.**, energy lost through respiration; ***F***, energy in faeces; ***U***, energy in urine; **Prod.**, total production. (From Penczak *et al.*, 1982.)

- the culture of carnivores;
- monoculture production systems.

The laws of thermodynamics, which describe how energy behaves, state that as organic matter is converted from one form to another it loses a proportion of its energy as heat (Odum, 1993). Thus, when a fish consumes algae it will use a proportion of the energy contained in the algae to fuel its metabolic processes, during which some of the energy is lost. Energy cannot be recycled, and the energy is lost from the ecosystem as heat and can no longer provide a source of metabolic energy. If that fish is then consumed by another fish, the same rules apply. Thus, the more transformations there are in the food chain, the more the energy originally in the algae is dissipated. The feed inputs to the trout farm in this example have probably been through two or three transformations before they reach the fish farm, and will undergo another to facilitate the growth of the trout. In comparison, the culture of tilapia in fertilized ponds, for example, is based on a much shorter food chain since the fish feed directly on phytoplankton and zooplankton; the ecological efficiency is therefore greater.

A further source of inefficiency, in relation to food production, is the loss of energy and nutrients from the system as waste: 80% of the carbon (which is a crude measure of energy content), 75% of the nitrogen and

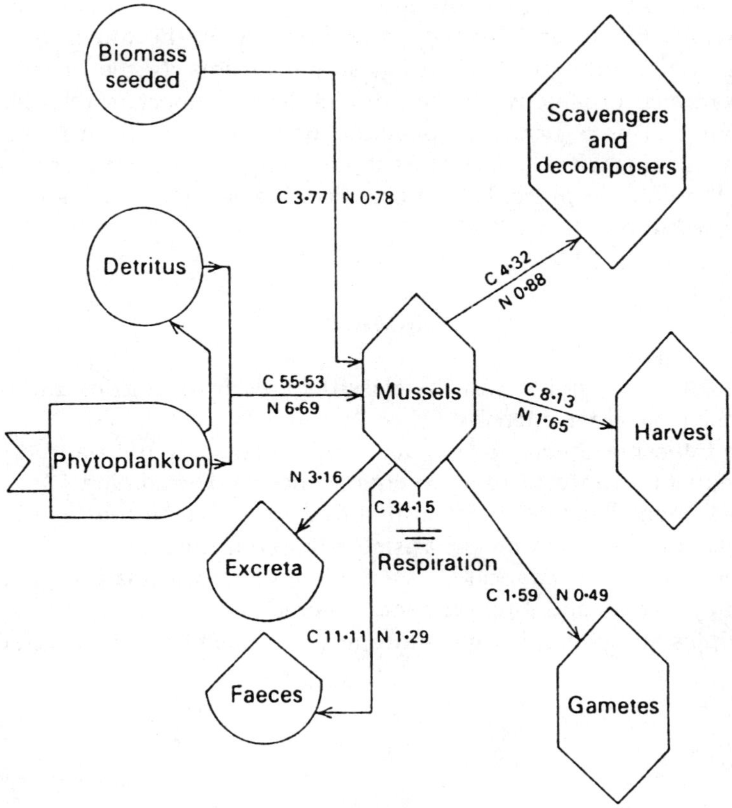

Figure 1.13 Flow of C and N through a mussel farm in Ireland. (From Rodhouse *et al.*, 1985.)

78% of the phosphorus. These wastes represent the potential for a whole new production process. Natural ecosystems are very complex, and this stems in part from the multitude of ways in which organisms have evolved to capture and make use of the energy that is contained in or excreted by other organisms. Monocultures are inherently inefficient because a single species will not be able to utilize fully all the energy available in the ecosystem; therefore polycultures or integrated production systems offer a much better model with regard to ecological efficiency.

(b) The bivalve farm as a system

The mussel is one of a variety of bivalves cultured around the world. With the possible exception of the oyster, it is the most technologically developed form of bivalve culture. Figure 1.13 depicts the flow of energy and bioelements through a mussel farming system (Rodhouse *et al.*, 1985), which has essentially the same elements as the intensive trout farming system in Figure 1.12. It is also responsible for the production of concentrated organic wastes but differs in one important respect: the food source. This is primarily plankton which has not been manufactured by humans with the use of fossil fuel energy or polluting processes, and it therefore represents a more sustainable system, despite the familiar waste problem.

SUMMARY

- Aquaculture systems can be classified along a continuum from extensive to super-intensive.
- More intensive aquaculture systems are ecologically unbalanced, and consequently can contribute to environmental degradation.
- Environmental management is required to minimize the severity of impacts on the environment arising from aquaculture.
- Systems ecology is advocated as a framework for the analysis of aquaculture systems and environmental interactions.
- Examples are given of aquaculture in the context of systems ecology.

CHAPTER 2

Environmental impact of aquaculture

2.1 INTRODUCTION

The environmental impact of aquaculture is seen in a variety of ways and includes user conflicts, alteration of hydrological regimes, introduction of exotic species to the wild and pollution of water resources. This chapter concentrates on the impact of wastes from aquaculture systems on the aquatic environment.

To put the extent of that impact into perspective, aquaculture must be compared with other industries that produce wastes with the potential to pollute. In most industrialised countries, modern methods of agriculture account for far more pollution of freshwaters and coastal seas than intensive fish farming. Without drastically changing agricultural farming practices this situation will not change in the forseeable future. For example, fish farming in Denmark accounts for less than 3% of the annual biological oxygen demand (BOD) on freshwaters, less than 1% of the total nitrogen loading and less than 2% of the total phosphorus loading (FES, 1992). In this respect the environmental impact from fish farming waste is small in comparison to the potential impact of agricultural wastes. In addition domestic sewage produces the greatest amount of waste phosphorus, which is known to have a substantial effect on freshwater aquatic environments (Table 2.1).

Data from other European countries indicate a very similar situation, although some caution must be exercised in the interpretation of data of such a generalized nature (differences in reporting and analytical methodologies can make them difficult to compare). Table 2.2, for example, shows that aquaculture in Scottish freshwaters contributes less to the annual phosphorus loading on Scottish freshwaters than forestry, and both contribute significantly less than the discharge from domestic sewage treatment systems.

Wastewater from intensive aquaculture operations also differs substantially from sewage effluents. The latter contain considerably greater amounts of organic matter, whilst aquaculture wastewaters are

Table 2.1 Sources of pollution in freshwaters in Denmark: Loading is in tonnes per annum; figures in parentheses are percentage of total (from FES, 1992)

Source	*BOD*	*Total N*	*Total P*
Agriculture	160,000 (66)	260,400 (89)	4,400 (29)
Domestic sewage	24,090 (10)	25,000 (9)	7,200 (47)
Industry	50,183 (21)	5,000 (2)	3,400 (22)
Fish farming	8,645 (< 3)	1,715 (< 1)	210 (< 2)

Table 2.2 Estimated annual phosphorus additions to Scottish freshwaters (from NCC, 1990)

Source	*Total*	*P addition rate* ($kg\ P\ y^{-1}$)	*Total P addition* ($t\ P\ y^{-1}$)
Aquaculture	(production) 4 300 t*	12 t^{-1} fish production	53
Atmospheric deposition	(area) 87 000 ha	0.12 ha^{-1}	10
Forestry	(area) 7.7×10^5 ha	0.1 ha^{-1}	77
Population	(people) 5.1×10^6	0.1 per person	510
Total P load			650

* Combined rainbow trout and salmon smolt production for 1989

more comparable to natural waters in terms of their nutrient composition (Table 2.3). However, whilst the concentration of pollutants might be small, over a period of time this can amount to quantities great enough to cause changes in the functioning of ecosystems.

This ecosystem change, or environmental impact, is normally seen on a local scale. Wastewater discharge from a fish farm into a relatively small river may cause oxygen depletion and community changes for only a short distance downstream, with the ecosystem returning to

Table 2.3 The quality of fish farm effluent, river water and domestic wastewater ($mg\ l^{-1}$) (from Warrer-Hansen, 1982)

Factor	*River water*	*Fish farm effluent*	*Domestic wastewater*
BOD	1.0–5.0	3.0–20	300
Total N	1–2	0.5–4	75
Ammonia N	no data	0.2–0.5	60
Total P	0.02–0.10	0.05–0.15	20
Suspended solids	no data	5–50	500

normal at some distance from the farm. A greater environmental danger is where aquaculture development becomes concentrated at a regional level. The effects of individual farms can be additive, and can be detected at the ecosystem level, in contrast to the more localized effects caused by a single discharge (Chapter 4).

Runoff from arable farming, the main source of ammonia-N, is difficult to quantify and detect. However, the effects of individual land-based fish farms can be detected and measured easily because of the point-source nature of their discharge. This has made the development and enforcement of legislative measures for fish farming relatively easy in comparison with those of the agriculture industry. It also provides the fish farmer with an advantage over the arable farmer since fish farm wastewaters can be treated and pollution minimized.

Most of the investigations into the waste output from intensive fish farming have concentrated on salmonid farms. The first major survey was undertaken in an *ad hoc* way for an EIFAC (European Inland Fisheries Advisory Commission) conference in 1981 (EIFAC, 1982). Much of the information presented was based on data from the late 1970s, and is still often quoted today. A similar survey was conducted in 1987 (EIFAC, 1988). However, since both surveys were undertaken considerable advances have been made in farm management and feed technology, and in the case of well managed farms the waste output is far smaller than that of 10 or 15 years ago.

2.2 AQUACULTURE WASTES

The most significant effect of aquaculture wastes is in increasing the nutrient concentrations in natural waters (hypernutrification) and so causing an increase in plankton and microbial populations (eutrophication). The source of these nutrients is mainly wasted fish feed. When the fish are fed not all the food presented to them is taken; some falls to the bottom uneaten. Various nutrients leach out of the feed pellet and contribute to the load of dissolved nutrients in the water. The pellet itself remains largely intact, to be eventually decomposed by microbes or eaten by wild fish or invertebrates.

Of the food that is eaten, a proportion of it is indigestible to the stock. This is excreted as solid faecal waste, which adds to the nutrient loading in the same manner as waste feed. Some is absorbed by the body and used for metabolic processes and for tissue growth. This process produces nitrogenous wastes, which are released into the water in a dissolved form as ammonia-N.

In addition to these organic wastes, residues of any chemicals or drugs used in the farming process will enter the water column or sediment. The impact of these on the environment is detailed in Chapter 3.

Thus, the wastewater from intensive aquaculture systems will contain:

- particulate solids (faeces and waste feed);
- dissolved metabolic wastes;
- dissolved nutrients from feeds and faeces;
- biocide and pharmaceutical residues.

2.2.1 Solids

Table 2.4 indicates the results of various investigations into the solid waste loadings from intensive freshwater fish farms. Even in this limited survey it is obvious that the amount of waste produced varies considerably, from 110 to 2153 kg t^{-1} fish produced in the examples presented here. This information is of limited use in environmental assessments for developing predictions of waste loadings from new aquaculture developments (Chapter 4) but it does emphasize the importance of good

Table 2.4 Solid waste loadings from freshwater salmonid fish farms, showing the amount of dry solids per tonne of fish production and in relation to stock biomass and feed input (modified from NCC, 1990)

Species and system	*Waste loading*
	kg t^{-1} fish produced
Survey of UK freshwater trout farms (EIFAC, 1982)	1350
Survey of UK freshwater trout farms (EIFAC, 1988)	546
National surveys (EIFAC, 1988)	
Austria	944
Finland	110–520
Spain	530
Greece	2153
Rainbow trout – ponds/tanks (Warrer-Hansen, 1982)	550
Rainbow trout – ponds/tanks (Warrer-Hansen, 1989)	300
Rainbow trout, 250 g fish in cages (Phillips, 1985)	289
Atlantic salmon smolts, FCR = 1.5 : 1 (NCC, 1990)	472
	g kg^{-1} food per day
Rainbow trout, cages (Merican and Phillips, 1985)	389.5
Rainbow trout, raceways (Makinen, 1983)	80–120
Rainbow trout, tanks (Clark *et al.*, 1985)	40–90
	g kg^{-1} fish per day
Rainbow trout, cages (Merican and Phillips, 1985)	5.49
Rainbow trout, raceways (Makinen, 1983)	110–520
Rainbow trout, tanks (Clark *et al.*, 1985)	0.8–0.94
Trout, Denmark, mainly earth ponds (EIFAC, 1988)	88
Mixed salmon, sea trout, rainbow trout, char, in tanks (Bergheim *et al.*, 1984)	4.8
Mixed salmon, rainbow trout, in tanks (Bergheim *et al.*, 1984)	3.3
Salmonids in tanks (HRPB, 1987)	6.4

farm design and good management. Design and management are the main factors in the variability evident in Table 2.4. Tanks, earth ponds and floating cages all have different hydrological regimes, which affects waste loading on the environment. Feeding is perhaps the main managerial factor, the type of feed and the feeding strategies used being of major importance in determining waste loading.

A factor that is often overlooked is that of temporal variations in waste loading. These are related to feeding periodicity, tank cleaning and pond harvesting operations, and seasonal changes in stock biomass on the farm (Poxton, 1991). Figure 2.1 illustrates the seasonal changes in stock biomass, feed input (related to stock biomass and water temperature) and waste solid loads for an Atlantic salmon smolt production unit in Scotland (NCC, 1990). The monthly waste loading varies from less than 50 kg in December to almost 400 kg in August and March.

The settling velocity of particulate waste in aquaculture wastewaters is important and has implications for the design of settling ponds and for the location of floating aquaculture cages. Figure 2.2 shows the settling velocity of trout faecal solids, based on data from a trout hatchery. As the fish become larger the faecal solids also increase in size (for fish above about 5 g in weight) and settle more quickly. In fact, species, individual size, proportion of waste feed and other factors will all influence the settling characteristics of particulate matter. It is difficult, therefore, to give any one value of settling velocity to characterize the solids in aquaculture wastewaters. As there will always be a great range of settling velocities in every situation, such data will be site specific.

2.2.2 Dissolved nutrients

The dissolved nutrients of most concern with regard to the environmental impact of aquaculture are nitrogen and phosphorus, in their various chemical forms. These are important nutrients for plants in the aquatic environment, and their increased availability in natural waters can significantly affect the balance of natural ecosystems. Whilst not generally regarded as an ecologically important nutrient in the same way as nitrogen and phosphorus, sulphur also can be important with regard to the environmental impact of aquaculture.

(a) Phosphorus

In order to appreciate the impact of phosphorus releases from aquaculture, the dynamics of phosphorus in the aquatic environment must be understood. This is discussed in the following section, and the implications of this are enlarged upon in Case 2.3, later in this chapter.

Of the total phosphorus entering a lentic aquatic system (a lake or a pond) the vast majority ends up bound in the sediment – in a highly

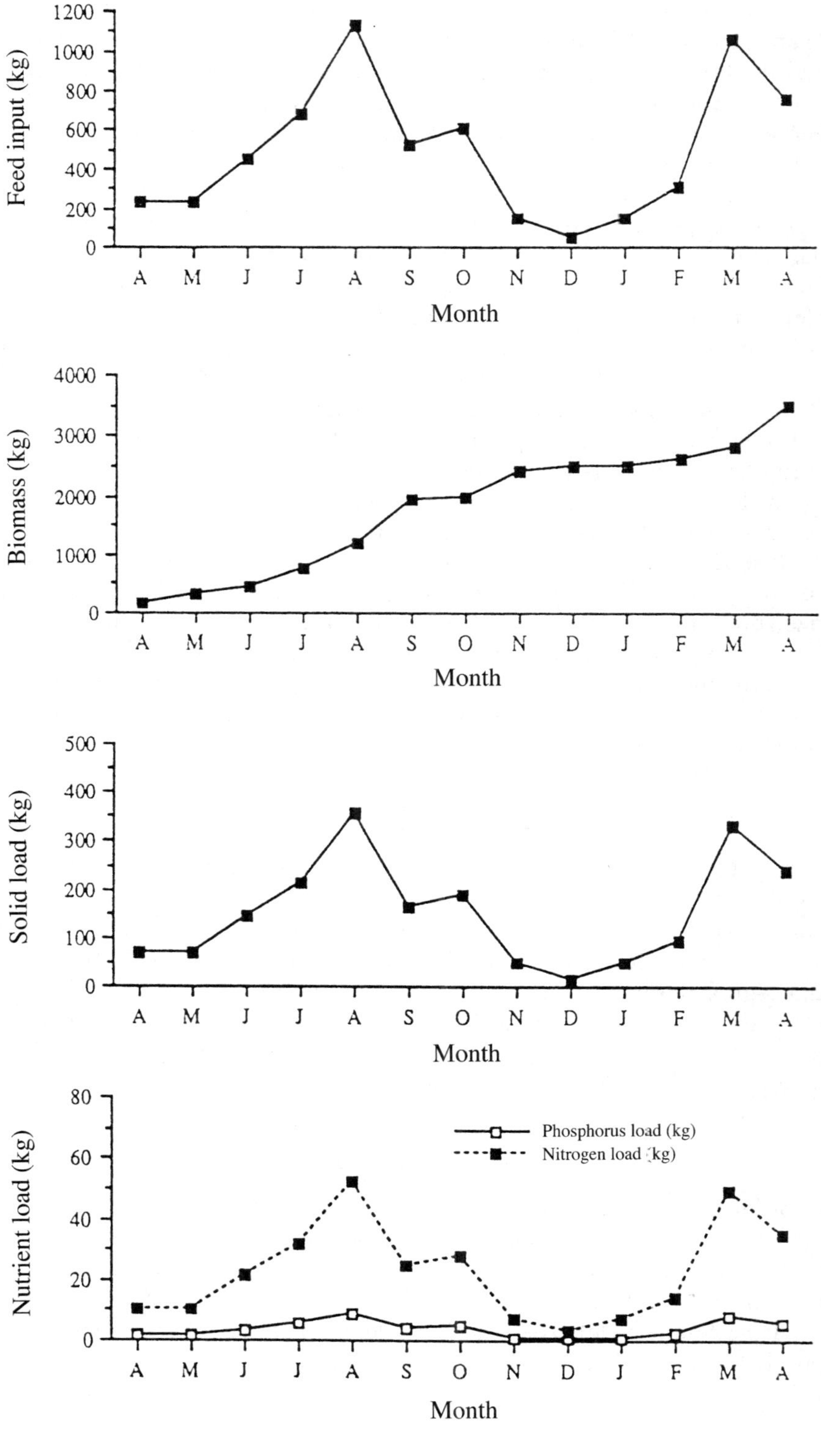
Feed input (kg)
1200
1000
800
600
400
200
0
A M J J A S O N D J F M A
Month
Biomass (kg)
4000
3000
2000
1000
0
A M J J A S O N D J F M A
Month
Solid load (kg)
500
400
300
200
100
0
A M J J A S O N D J F M A
Month
Nutrient load (kg)
80
60
40
20
0
Phosphorus load (kg)
Nitrogen load (kg)
A M J J A S O N D J F M A
Month

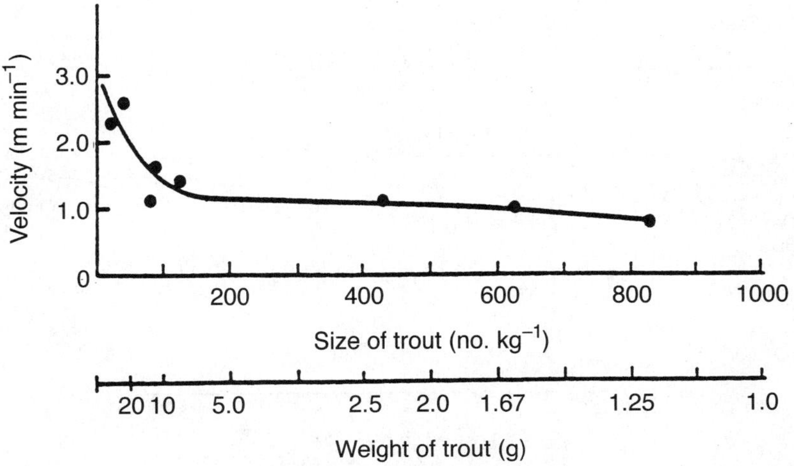

Figure 2.2 Settling velocities of intact excreta of rainbow trout. (From Warrer-Hansen, 1982.)

insoluble form under normal conditions. Soluble P is either taken up by plants for growth, or reacts with Fe^{3+}, Al^{3+}, Ca^{2+} or soil colloids, in which form it is tightly bound (Boyd, 1995).

Studies on fish pond soils (for static-water ponds) elicited the phosphorus budget illustrated in Figure 2.3. The two most important pools of P in the system were the fish and the pond soil (Boyd, 1995). When fish were harvested the P contained within their biomass was removed from the system. Over half the input P became bound in the soil in a relatively insoluble form. It did not then become completely unavailable to the pond water and plants, because the dynamics of the phosphorus cycle are such that an equilibrium is maintained between the soil and the water, but this equilibrium is highly skewed in favour of the soil. Measurements of total P in the different pools in a freshwater pond soil/water system showed that 99.81% of the P was bound in the soil.

The ability of soils to adsorb phosphorus is substantial, but even so it is possible to saturate a soil with phosphorus and greatly diminish its adsorption capacity (Boyd, 1995). Furthermore, soils rich in organic matter have a lesser capacity for phosphorus adsorption. The mineral particles, to which phosphorus is bound, make up a smaller proportion

---

Figure 2.1 Atlantic salmon (*Salmo salar*) smolt production showing seasonal changes in fish biomass and feed input and estimated waste loadings for the production of 100 000 S1 smolts (at 40 g per smolt). The calculations assume an FCR of 1.5 : 1 with 10% wasted feed and 50 000 potential S2 smolts graded and removed in October of the first year (i.e. a 33% S1 : S2 ratio out of 150 000 fry). (From NCC, 1990.)

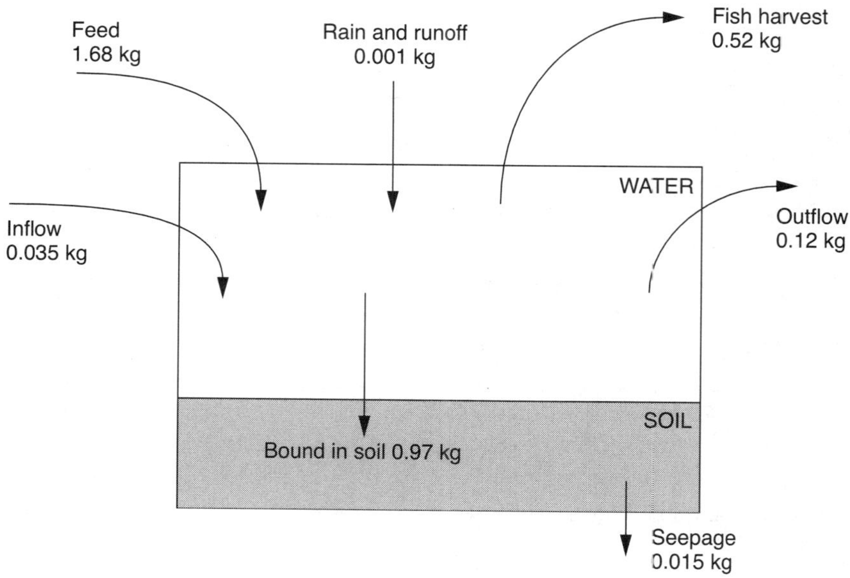

Figure 2.3 Phosphorus budget for a 0.04 ha channel catfish pond at Auburn, Alabama, USA. (From Boyd, 1995.)

of the soil, and the Fe^{3+}, Al^{3+} and Ca^{2+} ions in solution in the soil pore-water (i.e. that water lying between the soil particles, in the interstitial spaces) are largely complexed by organic compounds.

Phosphorus is strongly bound to mineral particles only under oxygenated conditions (Boyd, 1995). The situation changes dramatically if the water overlying the soil becomes anoxic. Under anaerobic conditions insoluble iron (III) phosphates are converted to soluble iron (II) phosphates. The soil pore water is often anoxic and highly reduced, especially if the soil is rich in organic matter. When phosphate and ferrous ions diffuse from the soil into the aerobic zone at the mud–water interface, they quickly precipitate as iron (III) phosphate compounds and fall back to the soil surface. If the overlying water is anoxic, then this phosphate remains in solution.

Some of the different measures of P in the aquatic environment are explained in Box 2.1.

(b) Nitrogen

Nitrogen occurs in aquacultural ecosystems in either inorganic or organic form. Organic inputs are manures and feed. Inorganic inputs of nitrogen are:

- chemical fertilizers;
- nitric acid in rainfall;
- existing inorganic nitrogen in the water supply.

BOX 2.1
MEASUREMENT OF PHOSPHORUS

Phosphorus ionizes in water to orthophosphate, of which there are a number of different forms that exist together in equilibrium ($H_2PO_4^-$, HPO_4^{2-}, PO_4^{3-}). The relative proportion of each in a solution at any one time depends on pH.

Much of this soluble orthophosphate is available to plants as a nutrient. Other forms of phosphorus present in natural waters include organic phosphorus, which can be degraded to orthophosphate by microbial activity, and phosphorus present as a component of particulate organic and inorganic matter.

There are various measures of phosphorus content used in organic chemistry, the most common being total phosphorus, total dissolved phosphorus and dissolved reactive phosphorus.

Total phosphorus includes chemically bound and freely available phosphorus. It is usually determined by an acid digestion method, which frees the phosphorus from organic and inorganic complexes.

Total dissolved phosphorus is measured using filtered water samples, and is a measure of the total remaining phosphorus (as orthophosphate). Filtering, with a specified grade of filter, removes much of the particulate phosphorus (which is largely insoluble in water), although very small particles, including bacteria and some protozoa, will be included in the measured value.

Dissolved reactive phosphorus is also determined from prefiltered water samples, and the test is performed over a limited period to ensure that only dissolved and loosely bound orthophosphate is measured. Plants are able to utilize only soluble orthophosphate. Dissolved reactive phosphorus, therefore, is the most relevant measure of the amount of phosphorus available as a plant nutrient in a freshwater system.

Organic nitrogen is decomposed by microbes to ammonia (NH_3), a process referred to as mineralization of organic nitrogen. Nitrification is a two-stage process by which ammonia-N is converted (oxidized) to nitrate, by chemo-autotrophic bacteria (Box 2.3). These are bacteria that synthesize organic matter from carbon dioxide, as do plants, but that derive the energy for this from organic compounds rather than from solar energy. Under anaerobic conditions nitrate may be converted to nitrogen gas, a process called denitrification.

Inorganic nitrogen exists in various forms, which are explained in more detail in Box 2.2.

BOX 2.2
CHEMICAL FORMS OF NITROGEN AND THEIR SIGNIFICANCE

With regard to aquaculture and environmental pollution, four forms of nitrogen are of significance: two forms of ammonia, and nitrite and nitrate.

Ammonia

The total ammoniacal nitrogen in water consists of:

- NH_4^+ – ionized ammonia (ammonium)
- NH_3 – un-ionized ammonia (UIA).

These two forms are in equilibrium in water, the balance of the equilibrium depending upon the pH and temperature of the water. At lower pH (more acidic water) the percentage of un-ionized ammonia is less than in more alkaline waters. The un-ionized fraction is more toxic to aquatic life. The following have been proposed (NCC, 1990) as guidelines for safe concentrations of un-ionized ammonia in fish culture waters, though there is considerable variation in tolerance between species (Poxton, 1991):

- salmonids: 0.002 mg NH_3 (UIA) l^{-1}
- non-salmonids: 0.010 mg NH_3 (UIA) l^{-1}
- marine fish: 0.05 mg NH_3 (UIA) l^{-1}

Nitrite

Nitrite (NO_2) is an intermediate product in the bacterial oxidation of ammonia to nitrate (NO_3) – the process of nitrification. In most natural waters and aquaculture systems nitrite is present only in very low concentrations. However, in recycle systems it can occasionally reach toxic levels where the biological filtration system is not working efficiently (Poxton and Allouse, 1987).

Nitrate

Nitrate (NO_3) is the end product of nitrification. Nitrate is not toxic to fish unless present at very high levels (> 100 mg l^{-1}) (Poxton, 1991), though there may be variations in tolerance between species. In recycle systems, it is eliminated from the culture water by continuous water exchange (about 5% per day). Nitrate is an important nutrient for algae in marine waters.

BOX 2.3
NITRIFICATION AND DENITRIFICATION

Nitrification is a two-stage process. Stage one is mediated by the bacterium *Nitrosomonas*, which oxidizes ammonium to nitrite (Lawson, 1995). Thus:

$$NH_4^+ + 1\tfrac{1}{2}O_2 \rightarrow NO_2^- + 2H^+ + H_2O$$

Stage two is mediated by *Nitrobacter*, which oxidizes nitrite to nitrate. Thus:

$$NO_2^- + \tfrac{1}{2}O_2 \rightarrow NO_3^-$$

Denitrification

Under anaerobic conditions nitrate may be reduced eventually to nitrogen gas, a process known as denitrification (Boyd, 1995). Again these reactions are carried out by micro-organisms. They use the oxygen contained in oxidized compounds for respiration in place of atmospheric oxygen. Thus nitrate is reduced to nitrite, nitrite to hyponitrite ($H_2N_2O_2$), and hyponitrite by a variety of pathways to nitrogen gas (N_2).

Aerobic decomposition of ammonia-N to nitrate is the most common pathway, and results in the release of this important plant nutrient into the water. In sufficient quantities (especially in seawater) this may stimulate rapid phytoplankton growth. In freshwater the development of phytoplankton blooms depends more on phosphorus concentration.

(c) Sulphur

Sulphur is of some interest in aquaculture because of its importance in anoxic sediments. In aerobic conditions organic sulphur decomposes to sulphide, which is in turn oxidized to sulphate. Sulphate is highly soluble in water, and so gradually disperses from sediments. Sulphide oxidation is mediated by micro-organisms in the sediment, though it can occur by purely chemical processes (Boyd, 1995).

Under anaerobic conditions, sulphate may be used in place of oxygen in microbial metabolism. This process leads to the production of hydrogen sulphide gas (H_2S) (Box 2.4). This often emerges from anoxic sediments under fish cages. It is highly soluble in water and has been suggested as the cause of gill damage and other ailments in fish (Beveridge, 1987).

BOX 2.4
HYDROGEN SULPHIDE PRODUCTION

Microbial production of hydrogen sulphide gas is characteristic of highly organic, anoxic sediments under fish cages. The H_2S is produced by a series of microbially mediated reductions (Boyd, 1995). Thus:

$$H_2SO_4 + 2H^+ \rightarrow H_2SO_3 \text{ (sulphite)} + H_2O$$

$$H_2SO_3 + 2H^+ \rightarrow H_2SO_2 \text{ (sulphoxylate)} + H_2O$$

$$H_2SO_2 + 2H^+ \rightarrow H_4SO_2 \text{ (sulphur hydrate)}$$

$$H_4SO_2 + 2H^+ \rightarrow H_2S + 2H_2O$$

2.3 ENVIRONMENTAL CONSEQUENCES OF HYPERNUTRIFICATION

The wastes from aquaculture systems enter the surrounding ecosystem and contribute to ecological processes. Studies undertaken in the last 15 years have enabled generalizations to be made about what the effects of these additions to aquatic ecosystems might be. It is not possible to give specifics, because each case of interaction between aquaculture and the environment is unique. Even a limited survey of the scientific literature will reveal the great variability in reported waste loadings from aquaculture and their environmental effects. What can be done is to provide examples that illustrate more specific cases and problems and give a deeper insight into the mechanisms at work.

The environmental impact of chemicals is dealt with separately in Chapter 3.

2.3.1 Oxygen depletion

Most of the environmental consequences of aquaculture are related to the addition of substances to the water. There is one notable exception to this pattern; that is, the depletion of dissolved oxygen. This is caused by:

- the respiration of intensively held animal stock;
- increased microbial respiration in the sediments and water column resulting from the release of organic carbon from aquaculture systems.

These pathways are each measured differently, and are consequently referred to in different ways. Dissolved oxygen depletion by respiration of the cultured animals is determined by direct measurement of dissolved oxygen concentration (DO) in the water column. Dissolved oxygen depletion by microbial respiration is measured by the determination of the biological oxygen demand (BOD). BOD is most often measured by incubating a sample of wastewater at 20°C for 5 days (BOD_5). The initial and final dissolved oxygen concentration is then determined and the difference between the two results indicates the relative oxygen-consuming properties of the water. This depends on:

- bacterial metabolic activity;
- the quantity of organic material present;
- algae and zooplankton metabolic activity.

The latter factor is usually eliminated by the filtration of wastewater samples prior to incubation. The bacterial metabolic activity, which causes the reduction in DO, depends in practice on the quantity of organic matter present as food. (Bacteria multiply very rapidly, so over a 5-day period the initial number of individuals is not likely to limit their total demand for oxygen during the incubation period.)

BOD is a commonly used measure of the severity of organic pollution, and may not seem directly relevant to oxygen depletion. However, the more severe biological effects of organic pollution are mainly due to the effects of oxygen depletion, examples of which are described in the following case studies.

CASE 2.1: WASTE FROM INTENSIVE SHRIMP FARMING

Shrimp farming in Thailand

In 1992 the shrimp farming industry in Thailand was at its peak. Farmers were harvesting up to 6 t of shrimp ha^{-1} per crop (national average of 2 t ha^{-1} per crop) with two to three crops per year. This was double the average yield of just 4 years previously. Nationally, the production figures were equally impressive. In 1986 only 18 000 t of cultured shrimp were produced. By 1989 this had risen by 500% to 93 000 t, and by 1992 the industry was producing 163 000 t of farmed shrimp. This was worth £600 million to £800 million in foreign exchange (Briggs, 1994; Dierberg and Kiattisimkul, 1996).

In 1992 there were about 16 500 separate shrimp farms in Thailand occupying 80 000 ha of coastal land – double the area used for shrimp production in 1988. About 85% of these farms were operated semi-intensively or intensively. Shrimp ponds are usually fertilized to stimulate natural live food production in the static ponds prior to the introduction of young shrimp (post-larvae) at a density of about 20 shrimp m^{-2}. As they grow they are given pelleted feed of high protein content, and water exchanges become more frequent in order to maintain the water quality. This is in contrast to the traditional extensive (low-intensity) system, in which ponds are stocked with wild-caught post-larvae at a density of less than 10 m^{-2}, and often in conjunction with fish. Annual yields of shrimp are commonly less than 200 kg ha^{-1}. Of these 16 500 shrimp farms, about 14 000 (80%) were small-scale operations, consisting of one or two ponds and of 0.5–1 ha in size (Briggs, 1994).

These impressive statistics masked the true state of affairs. The rapid expansion of the shrimp farming industry and the potential for high profits and short payback periods on investment capital encouraged short-term business goals. Sites were developed with little thought for the future, and consequently farms were often poorly sited and designed, and areas overdeveloped. Thus in 1990, along the northern coast of the Gulf of Thailand, one of the country's most important shrimp farming areas, many farms went bankrupt after only 2 years of operation. In the Samut Sakhon area, for example, 80–90% of the shrimp farms went out of business that year. Currently about 45 000 ha of once productive land lies derelict and of little use for any other purpose (Briggs, 1994).

In the more recently developed southern areas of Thailand where farms are better designed, there is still evidence that after only 4 years in operation similar problems may arise in the future. A survey in 1993 showed that farms which had been in production for only 3–4 years were experiencing a 24% decline in production. In the same survey 75% of farms were suffering disease problems, with an average loss in income of 22% per year as a result.

These problems have arisen largely as a result of overexploitation of the natural coastal resources and consequent poor water and sediment quality, leading to stress and disease and loss of production amongst the shrimp stocks.

Environmental problems

As with other forms of intensive aquaculture, pollution is directly related to the intensity of production, but in shrimp farming the effects of pollution on the life support environment of the shrimp are more direct. In pond or tank farming of salmonids, water exchange through the ponds is continuous and at a rate sufficient to remove the waste metabolites from the pond and to replenish the dissolved oxygen used for respiration before the fish are affected by these changes to their physicochemical environment.

Shrimp are farmed in semi-static ponds. The term semi-static is used here because water exchange is frequent and of a high volume, but it is not continuous. Thus, for much of the time the ponds are essentially closed environments and as the shrimp feed and respire they are exposed to increasing concentrations of waste metabolites, until the pond is next flushed. The self-polluting effect of overintensification in pond and tank systems is more direct and of more concern to the farmer than is the case with cage farming of fish.

The problem is further compounded by the discharge of waste metabolites and organic solids into the surrounding environment. In the central region of Thailand, south of Bangkok, some of the shrimp ponds are 4 km from the coast. Seawater is abstracted from tidal rivers or canals to maintain the water quality in the shrimp ponds. The wastewater is discharged into these same watercourses, along with that from numerous other farms (Chua *et al.*, 1989). As a result, the intensity of shrimp farming in these areas and the significant quantities of organic pollutants discharged into the coastal waters are overloading the ecosystem. Over an extended period a significant decline in the water quality can be detected in coastal waters in shrimp farming regions. This affects not only the marine environment but also the shrimp farming operations themselves.

The principal waste metabolite of shrimp is ammonia, which is highly toxic to most marine life, including the shrimp itself. Waste feed and faecal waste settles to the bottom of the pond, where it is decomposed by microbial activity. Typically the oxygen demand of these sediments is greater than that of all the shrimp in the pond (Briggs, 1994).

The result of this deterioration in the shrimp's physical environment is physiological stress. This lowers resistance to disease and so pathogens often flourish within the shrimp population, causing mortality and economic loss to the farmer. This can be seen in the high incidence of disease in intensive shrimp farms.

The scale of shrimp farm pollution could be large enough to cause large-scale environmental degradation and the series of events described here imply that this is indeed the case. However, if the situation is analysed more closely, the scale of the problem could be quantified. Nitrogen and phosphorus are the most important components of the organic waste from shrimp farming, as they are most likely to be in short supply in the aquatic environment and thus limit the growth of phytoplankton. Intensive shrimp farming in Thailand generates about 46 kg of organic waste ha^{-1} per day (Briggs, 1994). Some of this remains in the pond sediments, but about 1.2 kg N, 0.1kg P and 3.1 kg BOD per hectare of shrimp pond are discharged daily into the coastal waters. The Ranote/Hua Sui area in southern Thailand consists of a strip of land, only 40 km long and less than 2 km wide, containing 3520 ha of shrimp ponds (as of 1992). Approximately 4.1 t N, 0.4 t P and 11 t BOD are discharged daily into the coastal waters from this strip (Briggs, 1994). That amounts to approximately 1500 t of N, 146 t of P and 4000 t of BOD discharged annually along a 40 km stretch of coast. Organic pollution from shrimp farms has been linked to phytoplankton blooms in coastal waters, coral reef death, general decline in coastal water quality and eutrophication of wetlands (Chua *et al.*, 1989; Hopkins *et al.*, 1995).

In addition to organic waste, there is a significant problem caused by erosion of the ponds. The resulting sediment builds up on the pond bottom and is often flushed out when the ponds are harvested. Although this practice has been banned in Thailand, some farms still have central drains which operate on a continuous basis for pond cleaning, and which discharge directly to the receiving waters (Dierberg and Kiattisimkul, 1996).

The nutrient budgets for Thai shrimp ponds (Figure 2.4) indicate that 35% of nitrogen entering the pond is flushed into coastal waters. If the pond sediment is also flushed out with the wastewater, that total rises to 66%. For phosphorus, the impact of flushing sediments into the coastal environment is even more pronounced, with 84% of the input phosphorus being contained in the pond sediments (Briggs, 1994). It is important to remember, however, that as long as the sediments are not exposed to anoxic waters the phosphorus in them is largely unavailable as a nutrient.

Chemicals in the wastewater are also of concern with regard to their impact on the marine environment (GESAMP, 1991; Hopkins *et al.*, 1995). In particular, the indiscriminate use of antibiotics and other antibacterial agents has proved in the longer term to have been counterproductive. Resistant strains of pathogens have developed, making treatment harder and more costly, and ultimately increasing mortalities (Karunasagar *et al.*, 1994; Hopkins *et al.*, 1995). Hatchery-reared post-larvae, having been continually protected against disease by the prophylactic use of these agents, are much more likely to succumb to disease during the grow-out stage. Furthermore, antibiotic residues detected in harvested shrimp seriously affected sales of Thai shrimp to Japan in 1991 and 1992 and brought about a 27% reduction in the price of Thai shrimp on world markets almost overnight. There are concerns that in the marine environment residues of these chemicals may disrupt natural bacterial decomposition processes and alter the structure of marine microbial communities (Briggs, 1994).

CASE 2.2: IMPACT OF TROUT FARM WASTEWATER ON A STREAM ECOSYSTEM (Camargo, 1992)

Rio Tajuña is a limestone stream, typical of the Tajo Basin in the Gaudalajara region of central Spain. Its waters are hard and alkaline as a result of the dissolution of minerals from the calcareous bedrock, and on average flow at a rate of 0.6 $m^3 s^{-1}$ (this varies with season). The stream ranges from 2 to 6 m in width, and consists of a series of riffles (shallow, rocky sections) and deeper pools. About 13 km from the source of the river is a trout farm, producing about 75 t of trout annually. Water is abstracted from the stream, fed through a series of trout ponds, then discharged via a sedimentation pond further downstream (Figure 2.5).

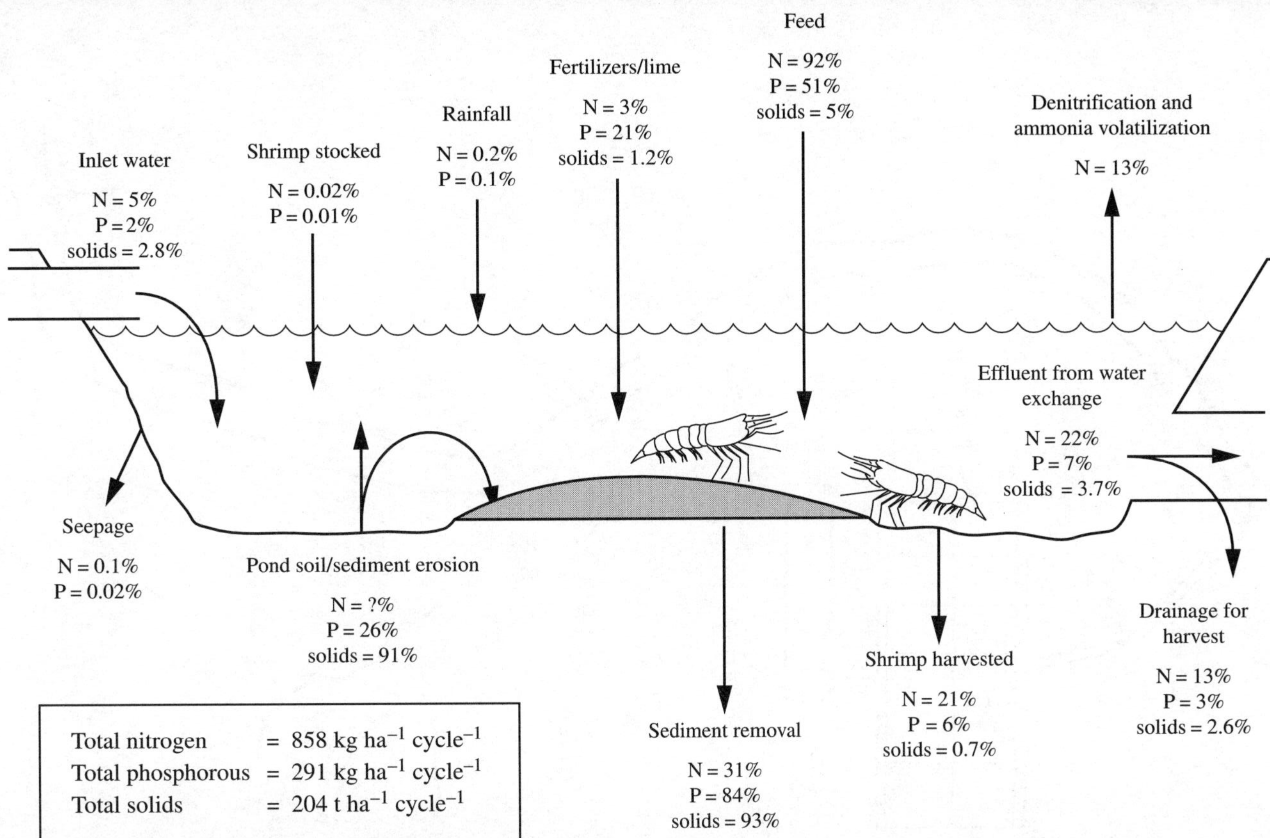

Figure 2.4 The sources and sinks of nitrogen (N), phosphorus (P) and solids in intensive shrimp ponds. Numbers refer to the percentage of the total either introduced to or exported from the pond, as indicated by the direction of the arrow. (From Briggs and Funge-Smith, 1994.)

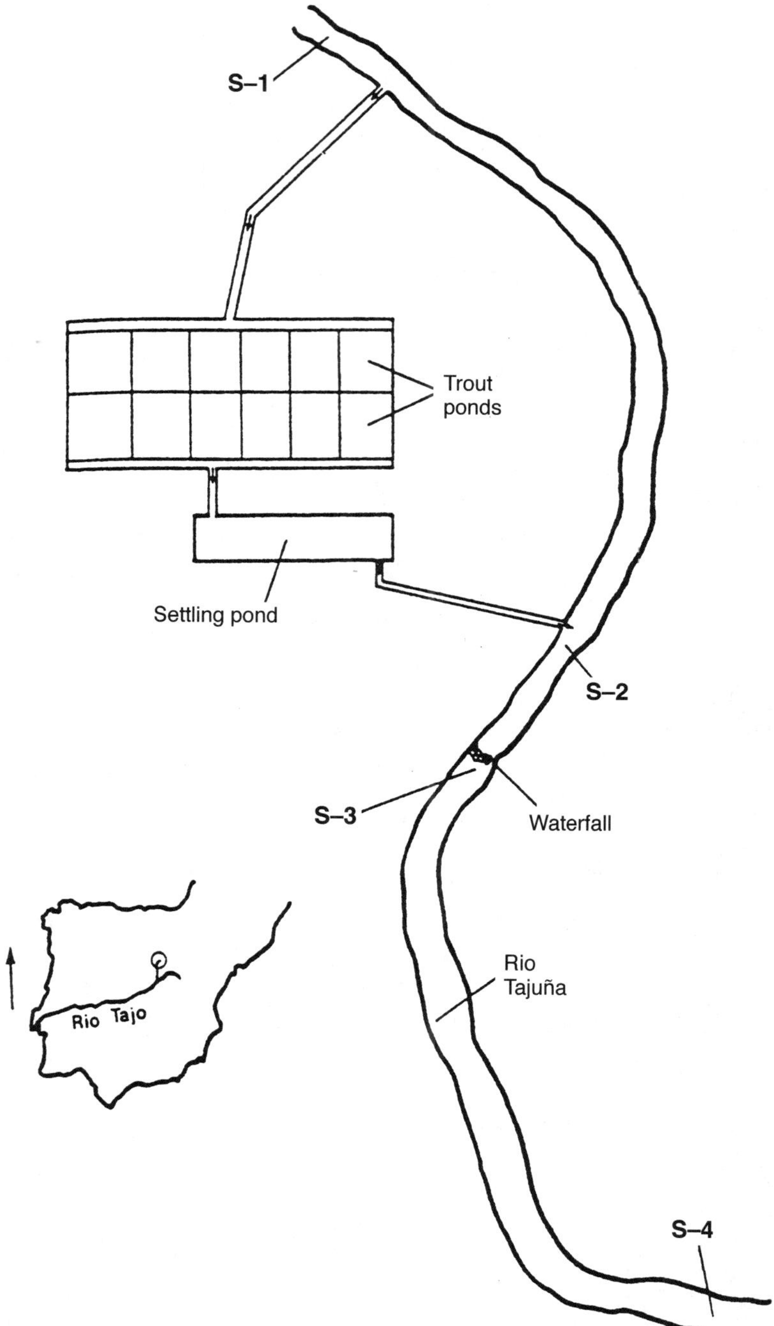

Figure 2.5 General diagram of a Spanish trout farm showing the location of sampling stations in the Rio Tajuña. (From Camargo, 1992.)

As this farm is representative of many of the trout farms in Spain (approximately 200), it was selected for a study to investigate the effects of its discharge on the benthic invertebrate community in the stream.

Benthic invertebrates have been investigated for a long time as part of ecological studies into the effects of pollution in freshwaters, because they are particularly sensitive to changes in their physicochemical environment. Some species are especially sensitive, whilst others are more tolerant of pollution. Thus, particular invertebrates have been identified as indicator species for different levels of pollution. Various biotic indices have been developed, based on different permutations of raw data collected by population sampling techniques, so that scientific evaluations can be made of surveys of benthic macro-invertebrate populations. Many of these were developed in the UK, the most familiar being the Trent Biotic Index (TBI) and Biological Monitoring Working Party (BMWP) score. In continental Europe the Saprobian system is more widely used (Magurran, 1988; Gray, 1989).

By using these indices it is possible to determine how much the different sections of a stream are affected by pollution. The use of indices is a faster alternative to expensive and time-consuming chemical analyses on a large number of water quality indicators (e.g. ammonia, nitrite and nitrate, orthophosphate, pH, dissolved oxygen, BOD, suspended solids, etc.). Also, these indices reflect the ecosystem quality over a long period, whereas water quality parameters reflect only the conditions at the time of sampling.

To assess the extent of the impact of the trout farm wastewater at the discharge point and downstream, four sampling stations were chosen in this case study. Thus, the investigation would not only shed light on how a typical effluent affected the stream community at the point of discharge, but also give some idea as to how far downstream any effect extended. From the outfall to the point at which the ecosystem returned to normal is referred to as the recovery zone.

The sampling stations were located above the abstraction point, to measure the 'normal' condition of the stream, and at 10, 150 and 1000 m downstream of the discharge point. Figure 2.6 shows some of the water quality measurements from a previous study at the same location, using the same sampling stations. The effect of the trout farm wastewater on the water quality in the stream at the outfall was considerable, and even 1 km downstream the concentrations of ammonia, nitrite and orthophosphate remain elevated. So, what effect did this deterioration in the water quality have on the benthic macro-invertebrate community and what was the environmental impact of the trout farm effluent?

Table 2.5 shows the change in value of a selection of biotic indices at the sampling stations in the Rio Tajuña. Note that all the indices exhibit a similar pattern. An example is the species richness index, which is a measure of the number of species present. This indicates that the number

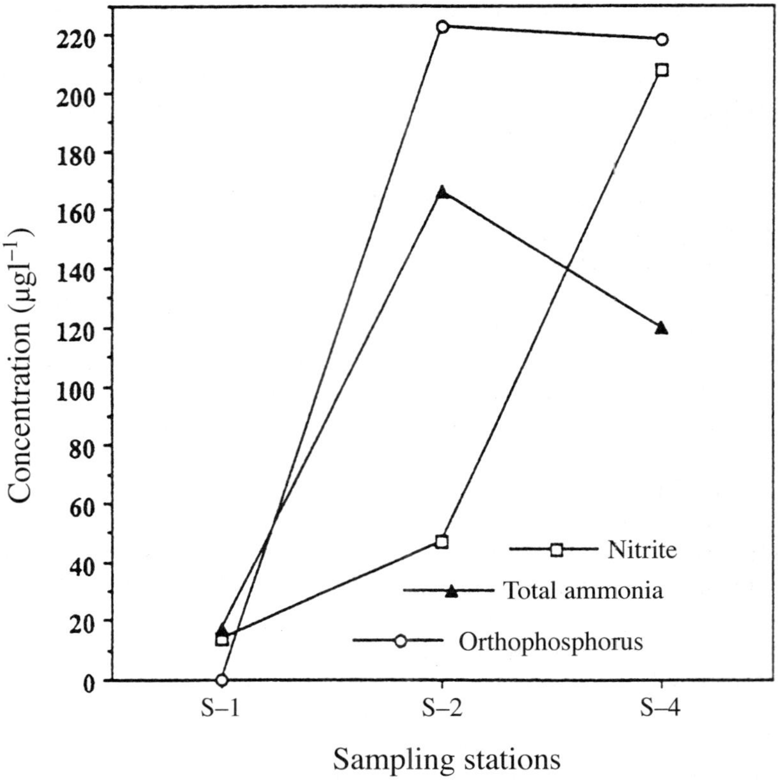

Figure 2.6 Mean concentrations (n = 3) of nitrite, total ammonia and orthophosphorus at S-1, S-2 and S-4 sampling sites on Rio Tajuña during July 1986. (From Camargo, 1992.)

of species present in the benthic macro-invertebrate community falls considerably at the outfall from the farm, and at a distance of 1 km downstream the species number has not returned to the value measured upstream of the outfall. It is interesting to note that the indices all indicate that the stream is less polluted at sampling station 3 than at station 4. This is attributed to the effect of the small waterfall just above the third sampling station and its positive effect on the physicochemical conditions in the stream, such as dissolved oxygen concentrations. However, the effect is only short-lived. The large quantity of organic sediment in the stream, which had settled from the trout farm wastewater, requires considerable quantities of oxygen for microbial decomposition and releases large amounts of ammonia back into the water as a by-product of this process.

A closer examination of the presence and ecology of the benthic macro-invertebrate species at each station reveals that species adapted to

Table 2.5 Values of biological indices at each sampling station (distance from the outfall in parentheses) in summer surveys (modified from Camargo, 1992); the location of sampling sites is indicated in Figure 2.5.

Biological index	*S-1 (0 m)*	*S-2 (10 m)*	*S-3 (50 m)*	*S-4 (1 km)*
Species richness	38	13	28	21
Trent biotic index	10	3	9	8
Chandler biotic score	2052	271	870	678
Biological Monitoring Working Party score	168	19	64	53
Average score per taxon	6.72	2.71	4.24	4.08
Overall quality rating	7	2	3.5	3

shallow fast-flowing streams (which have high dissolved oxygen concentrations and low sedimentation rates) have disappeared. They have been replaced by species more characteristic of slow-flowing lowland rivers with environments characterized by lower dissolved oxygen concentrations and high sedimentation rates (and often muddy substrates). At sampling station 2, in a deeper pool just below the waterfall, the bottom of the stream was covered with a thick deposit of anaerobic sludge. This had accumulated through the settlement of organic solids (waste fish feed and fish faecal wastes) from the trout farm. It was inhabited mainly by tubificid worms and certain chironomid larvae. These are red, having haemoglobin and other respiratory pigments (which take up oxygen very efficiently) in order to survive in conditions where oxygen is in such short supply.

CASE 2.3: FLOATING CAGE FARMING

Floating cage farming is most commonly used for salmonid-growing in marine and freshwaters, and the environmental impact of this activity has been studied comprehensively (e.g. NCC, 1990; Bergheim and Asgard, 1996). The findings of this research are just as applicable to other species cultured in this way, e.g. yellowtail, sea bass and sea bream, though differences in climate and in feeding practices must be taken into account.

Pollution from floating cage aquaculture can affect the environment via two mechanisms. Fish faecal waste and waste feed settles to the bottom of the lake or sea bed, and accumulates to form a thick, organically rich sediment. This affects the benthic community and can affect water quality. Also, respiratory wastes (primarily ammonia) and dissolved phosphorus (leached from waste feed and faecal waste) may affect nutrient flux in the water body, and thus the dynamics of plankton communities.

Studies on the impact of solid wastes

Loch Spelve is a small Scottish sea loch on the west coast island of Mull. In 1985 it was the subject of an investigation into the impact of the solid wastes from a salmon farm on the benthos of the loch (Brown *et al.*, 1987). This was a floating cage farm arranged in a block of six cages, 18 m wide and 30 m long in total, and supporting an adult salmon biomass of 35 t.

The mooring system for floating cages has an important impact on the area of sea bed affected by the waste solids from the farm. On a single-point mooring the block of cages is free to move in a complete circle, under the influence of tidal currents and prevailing winds. A multiple-point mooring system restricts the movement of the block considerably. A block on a single-point mooring will spread solid wastes over an area 20–40 times greater than a block of cages with multiple-point moorings, depending on the radius of the arc described by the former (Figure 2.7).

The salmon farm studied in Loch Spelve had multiple-point moorings and the movement of the block of cages was restricted to up to half its

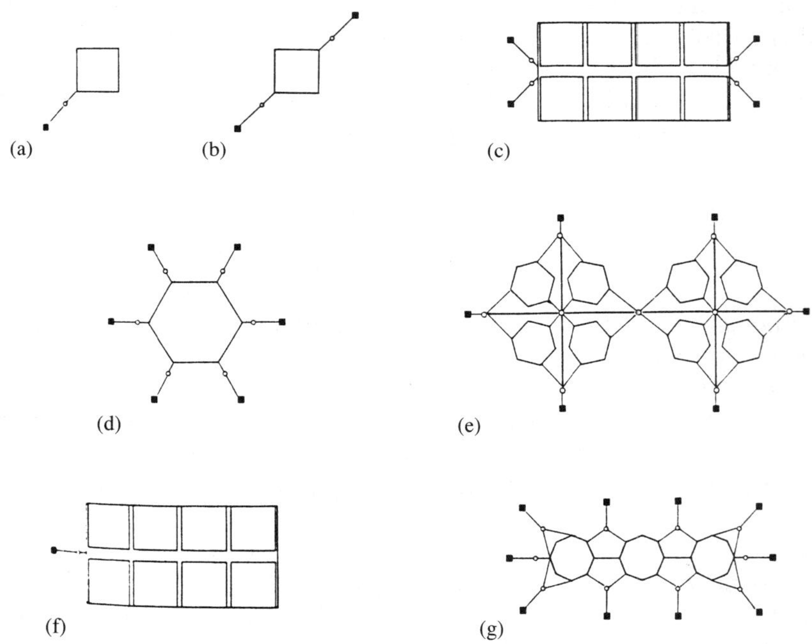

Figure 2.7 Examples of various cage groupings and mooring systems: **(a)** single cage, single-point mooring; **(b)** single cage, multiple-point mooring; **(c)** group of cages, multiple-point mooring; **(d)** giant cage mooring; **(e)** hexagonal cages moorings; **(f)** group of cages, single-point mooring; **(g)** octagonal cages moorings. (From Beveridge, 1987.)

length and half its width in either direction. Thus it covered an effective area of 2160 m^2 of sea bed. The cages were moored in water 20 m deep and the average current velocity in the vicinity of the cages was 0.037 m s^{-1}. The investigation itself had two main elements: a study of the physical and chemical nature of the sediments in the vicinity of the cages, and a study of the benthic communities (the benthos) around the site. The objective was to find out whether the presence of the cages had had a significant impact on the benthos, and how far from the cages was any effect discernible.

As mentioned previously, in studies of pollution in freshwaters benthic macro-invertebrates are used as indicators of the severity of pollution (by considering the composition of the community in respect of species number and abundance of each species). This technique also works in marine environments, though the indicator species are, of course, different. A reason for using bioindicators in pollution studies is that they reflect the long-term average in environmental conditions, or ecosystem health. Measurements of chemical attributes of the environment (such as water quality), on the other hand, reflect conditions only at the time of sampling. They are, nevertheless, a valuable way of building up a detailed picture of day-to-day processes and interactions.

Figure 2.8 shows the results of both aspects of the study – the presence of particular macro-invertebrate species being closely linked to the physical and chemical nature of the environment. Not surprisingly, the condition of the benthos and benthic sediments was worst directly under the cages. Here, the sediment was of very fine particle size and with a high carbon content compared with unaffected areas – it was typical of an organic sludge. Its colour was black, indicating anoxic conditions, and bubbles of hydrogen sulphide gas regularly erupted from its surface. No macro-invertebrates lived in this zone, and the only obvious sign of life was a mat of *Beggiatoa* sp., a bacterium that lives at the interface between sulphate-rich and oxygen-rich environments.

The next identifiable zone was between 3 m and 15 m from the cages. The biotic community of this zone was dominated by the polychaete worms *Capitella capitata* and *Scolelepis fuliginosa*. Although the sediments were still highly anoxic, their effect on the overlying water (i.e. reducing the dissolved oxygen concentration) was not as serious as in zone 1 (except in early summer, when sulphate bacteria were evident and outgassing of H_2S could be seen) and these worms were able to survive.

Zone 3 extended from 15 m to 120 m from the edge of the cages. The number of species present in this zone and their abundance in the sediments was much greater than for zone 2. *Capitella capitata* was still present in large numbers but was not totally dominant. It is possible that this zone, in comparison with unaffected sediments, may benefit from the increased availability of organic matter in providing an abundant food source for the benthos. Once enrichment has reached a level

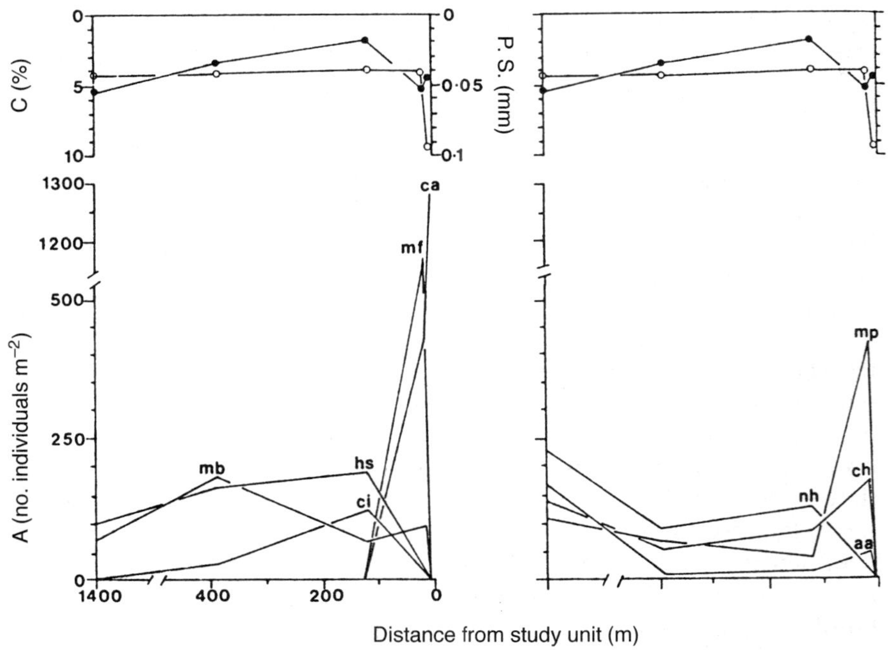

Figure 2.8 Data from a survey of the benthos under marine salmon cages. Plots against distance of: **A**, mean abundance of selected dominant species; **C**, mean sedimentary carbon content (○ carbon content, ● particle size); **P.S.**, mean particle size. For clarity, the species have been separated into two diagrams. **aa**, *Abra alba*; **ca**, *Capitella capitata*; **ch**, *Chaetozone/Tharyx* spp.; **ci**, *Cirratulus cirratus*; **hs**, *Heterocirrus* sp.; **mb**, *Mysella bidentata*; **mf**, *Mediomastus fragilis*; **mp** *Melinna palmata*; **nh**, *Nephtys hystricis*. (After Brown *et al.*, 1987.)

at which dissolved oxygen in the interstitial water (that between the sediment particles) and in the overlying water falls to very low levels, this benefit is lost. Conditions are such that organisms needing an aerobic environment cannot make use of the organic matter present. Instead, anaerobic microbial processes take over, and make the environment even less suitable (e.g. by the production of H_2S and methane gas) for a normal benthic marine community. It is worth noting, however, that this 'benefit' is not necessarily an advantage for the ecosystem. With enrichment at this level the ecosystem is less stable and less able to cope with small changes in the processes occurring within it. For example, in zone 3 it was noted that in early summer the sulphate bacterial activity increased, as evidenced by the outgassing of hydrogen sulphide and the growth of *Beggiatoa* sp. A small change in the environment – perhaps water temperature – had switched the sediment environment from a tolerable to an intolerable one for much of the benthic community.

BOX 2.5
REDOX POTENTIAL AND ENRICHED ORGANIC SEDIMENTS

The redox potential is a frequently used measurement in studies of the environmental impact of aquaculture (Boyd, 1995). It is a measure of the flow of electrons as oxidation or reduction reactions occur. Electrodes are available that can measure the redox potential of, for example, pond water or lake sediments. The higher the value (usually given in units of E_h), the more oxidized is the substance; and the lower the value, the more reduced is the substance. A highly reducing condition is usually indicated by a negative value for E_h, although different electrodes may be calibrated differently. pH and temperature can also affect the redox potential measurement, and so adjustment to equivalent values for pH 7 and 25°C may be necessary for valid comparisons to be made.

In the context of this book, redox potential is useful as an indicator of the degree of microbial activity in organically enriched sediments, thereby indicating the degree of organic enrichment. The causal relationship is that as the microbial population increases (which it will do in proportion to the availability of organic substrate for metabolism) its demand for oxygen increases. Reducing conditions develop as the oxygen concentration in the environment decreases, resulting in negative measurements of the redox potential.

At more than 120 m from the cages (zone 4) the benthos and sediment were considered normal for this site.

Thus, the environmental impact of solid wastes from floating cage aquaculture tend to be quite localized. The effects are also reversible, if the cages are moved elsewhere, and they can be lessened by appropriate management measures (Chapter 4).

Studies on the impact of dissolved wastes

Studies on the effect of dissolved cage aquaculture wastes, especially in marine environments, are difficult because of the diffuse nature of the pollutants. Its effects are therefore less localized and more long term than the effect of solid wastes.

There is sufficient published information to develop a generalized picture of the processes and their impacts on the environment (e.g. NCC, 1990; Bergheim and Asgard, 1996). Figure 2.9 illustrates the fate of nutrients derived from fish feed in floating cage farms. Studies have shown that about 60% of the carbon contained in fish feed eventually finds its

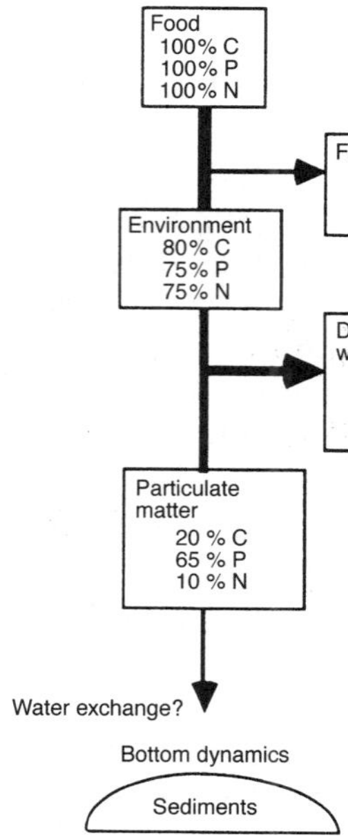

Figure 2.9 Illustration of the flow of bioelements through net pens in fish farms rearing salmonids; based on a dry feed conversion ratio of 1.5. (From Bergheim *et al.*, 1991.)

way, as a dissolved compound, into the aquatic environment around the fish farm. Similarly, 10% of the P and 65% of the N also enter the aquatic environment in dissolved form. In addition, microbial activity in the benthic sediments will release further amounts of C, N and P as dissolved substances (Bergheim *et al.*, 1991). More recent work (in Bergheim and Asgard, 1996) indicates dissolved waste loadings of 41% of input N and 18% of input P. Whether the difference between this and the data of Bergheim *et al.* (1991) is due to differences in feed quality, management practices or other factors is not clear.

An important point to note is that most waste phosphorus is bound up in the sediments, where it remains relatively unavailable to aquatic life. This is important in freshwater lakes, where phosphorus is normally the limiting nutrient to phytoplankton growth. If it were not bound up in the sediments, but remained available as a nutrient for phytoplankton, the eutrophication of freshwaters used for fish farming would be greatly accelerated. In the marine environment, nitrogen is normally the limiting nutrient and so the effect of intensive fish farming on eutrophication in the marine environment is potentially much greater. However, eutrophication of the wider coastal marine environment is difficult to attribute to one particular source, because of the extent of tidal and wind-induced mixing and mass water movements due to oceanic currents, etc.

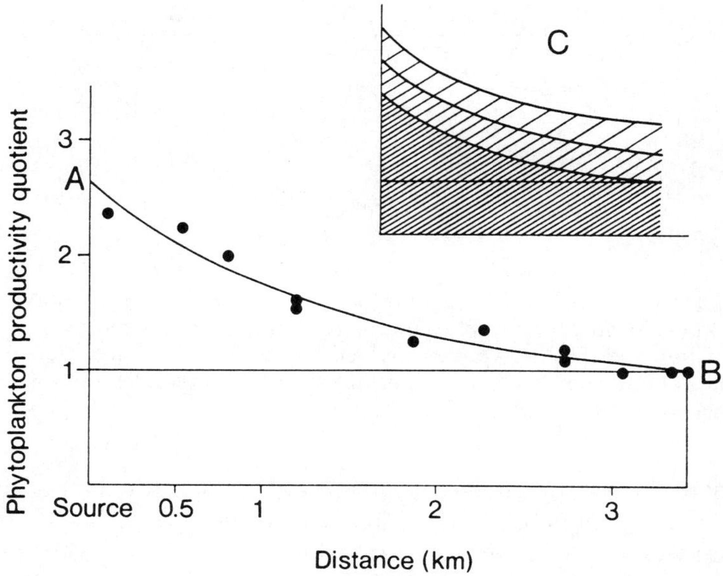

Figure 2.10 Mean primary productivity (^{14}C measurements) at various distances from fish farm **A** in Lake Längelmävesi, Finland, related to primary productivity of station **B**, 3 km from source A. Inset **C** shows expected increase of productivity level of the lake in addition to formation of a gradient. (From Persson, 1991.)

Freshwater lakes, in comparison, are more isolated environments with respect to water exchange and it is consequently easier to identify the sources and effects of pollution. For example, measurements were made of the level of primary production in Lake Längelmävesi in Finland (Persson, 1991). Figure 2.10 shows that primary productivity is much higher nearer to the floating cage fish farm located in this lake. Freshwater lakes in higher latitudes are normally oligotrophic; that is, they characteristically have a low concentration of nutrients in their waters, and consequent low levels of primary productivity. The ultimate effect of hypernutrification and eutrophication is to make the lake ecosystem unstable, resulting in plankton blooms and crashes, and associated large fluctuations of dissolved oxygen concentrations in the lake. The effect on the biotic community, adapted to stable lake ecosystems, is severe.

The degree of eutrophication attributable to fish farming in freshwater lakes in cold climates rarely approaches anywhere near this situation. In the majority of situations eutrophication of this nature is barely detectable. However, these oligotrophic lakes are very sensitive to changing nutrient status, and the intensity of fish farming in these ecosystems should be carefully regulated to ensure that the carrying

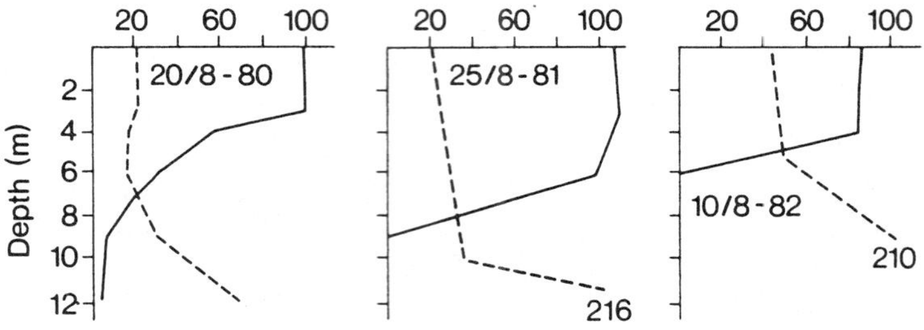

Figure 2.11 Effects of establishment of a rainbow trout fish farm on oxygen saturation (solid line, %) and phosphorus concentrations (broken line, $\mu g\ l^{-1}$) in Lake Gunneröd, Sweden, pre-eutrophied by municipal sewage. Due to anaerobic conditions in the deep water, rapid phosphorus release from the sediments was induced. (From Persson, 1991.)

capacity of the ecosystem (i.e. the amount of extra nutrients it is able to process without significant effect) is not exceeded.

A rare example of severe hypernutrification, at least partly due to intensive fish farming, occurred in Lake Gunneröd in Sweden (Persson, 1991). When a licence was given for a small floating cage trout farm in 1978, the lake was already in a eutrophic state as a result of municipal sewage discharge. The fish farm contributed to the already significant nutrient inputs, and gradually the condition of the lake deteriorated even further (Figure 2.11). Eventually the lake's hypolimnion (the deeper, cooler waters) became completely anoxic, thus restricting most aquatic life to the top 5–6 m of water, the epilimnion (Box 2.6). The depletion of oxygen in the deeper waters had an important consequence. Phosphorus, discharged into the lake by sewage disposal and the fish farming operations, had been building up in the sediments over a number of years. As mentioned previously, 65% of phosphorus discharge in fish farm waste is bound up in the sediments and this remains the case as long as the waters overlying the sediment remain oxygenated. However, if these waters become anoxic the resulting chemical changes release significant amounts of phosphorus into the water. The dissolved phosphorus is available for plant growth and thus eutrophication is accelerated. This becomes a self-sustaining process, illustrated in Figure 2.11 with respect to Lake Gunneröd.

The example of Lake Skärsjön, again in Sweden, is much less extreme but much more typical (Persson, 1991). A long-term study of the nutrient status of the lake and its ecology detected significant changes following the establishment of a fish farm in the lake in 1978 (Table 2.6 and Figure 2.12). The secchi disc depth, which provides a measure of phytoplankton density in this case, indicated a progressive increase in the plankton populations from summer to summer. Similarly the phosphorus concentrations in the lake were much greater than previous levels during the

BOX 2.6
LAKE STRATIFICATION

Warm water is less dense than cold water and, therefore, rises to the surface. It also resists movement into cooler, more dense water. The greater the difference in temperature, the more wind energy is needed to keep the lake waters fully mixed (wind-induced mixing keeps temperature and water quality fairly even throughout the lake).

In the summer in higher latitudes, the surface waters usually warm up quite rapidly. This is at a time when winds are usually light, so that little mixing occurs. The result is a stratification of water in the lake, with a distinct boundary called the thermocline. It is a zone of rapid temperature fall and represents a physical barrier to mixing. Once the thermocline has become established, the waters above it – the epilimnion – remain mixed and well oxygenated, whilst those below – the hypolimnion – are cut off from fresh oxygen supplies (Moss, 1980). If the oxygen demand in the hypolimnion is high the oxygen may be used up and the hypolimnion becomes anoxic. This may happen if, for example, it contained enriched organic sediments and had nutrient-rich waters in which plankton and bacteria could flourish. The resulting high BOD could lead to anoxia.

In the autumn as the ambient air temperature falls the surface water temperature gradually falls. As the temperature difference between the hypolimnion and epilimnion decreases and as the wind speeds increase, which they normally do in temperate latitudes at this time of year, then the thermocline breaks down and the lake is mixed to its full depth. This is the autumnal overturn. If the hypolimnion has become anoxic during the summer, the overturn can lead to massive mortality amongst the biotic community.

Deep tropical lakes, such as the rift valley lakes in East Africa, are often permanently stratified. Without a cold winter to cool the surface waters to a temperature similar to that of the hypolimnion, there is no overturn. This results in water in the hypolimnion being isolated from the atmosphere and under conditions of low light intensity. Without light the production of oxygen through photosynthesis is low or non-existent. If bacterial respiration associated with detritus or sediments continues, the amount of oxygen will decrease often to zero and the hypolimnion becomes anoxic (Moss, 1980). Under low oxygen conditions the sediment–water interface may undergo chemical changes that result in the release of inorganic ions such as FeII, MgII, and PO_4^{3-}, which in turn may affect the benthic community.

Table 2.6 Morphometric and hydraulic data for two lakes eutrophied by emissions from cage aquaculture (modified after Persson, 1991)

Factor measured	*Lake Skärsjön*	*Lake Gunneröd*
A (km^2)	3.1	0.18
Z_m (m)	22.5	13
Z (m)	9.0	4.3
T (years)	6.0	1.0
SQ ($l\ s^{-1}$, km^2)	14	10.3
Total P, before ($\mu g\ P\ l^{-1}$)*	5–7	23–28
Total P, after ($\mu g\ P\ l^{-1}$)*	14.1	45.0
Secchi depth, before (m)*	6.5–8.5	–
Secchi depth, after (m)*	3.5–6.5	–
Chlorophyll *a*, after ($\mu g\ l^{-1}$)	6.1	–

A = late area; Z_m = maximum depth; Z = mean depth, T = water renewal time; SQ = specific runoff.
*Selected variables indicative of trophic conditions include summer means or averages of total phosphorus (Total P), secchi depth and chlorophyll *a* concentration.

period that the fish farm was located in the lake. As well as these indications, the species composition of the phytoplankton pointed to a shift in the nutrient status of the lake from oligotrophic to mesotrophic. About 70 taxa (individual species or groups of closely related species) were recorded, including those typical of oligotrophic lakes and those indicative of more eutrophic conditions. This variety is typical of a transitional situation, where previously rare species are expanding and the original flora or fauna still persists.

The benthic fauna showed similar changes. Firstly, there was a change in the dominant species in the chironomid community from taxa less tolerant to pollution (*Tanytarsus* spp.) to the more tolerant *Chironomus anthracinus*. This species was in turn replaced by oligochaetes as the dominant benthic group, an indication of advancing eutrophication. The danger presented to the ecology of the lake prompted the closure of the fish farm in 1985. Subsequently the dissolved oxygen levels at the bottom of the lake remained between 3 and 4 mg $O_2\ l^{-1}$ but in 1986 it fell to 0.1 mg $O_2\ l^{-1}$ for a short period. Fortunately, there was no mass release of phosphorus from the sediments and the lake is now recovering.

Localized anoxia may still occur in waters less polluted than those described here. The BOD of enriched sediments directly under fish cages can deplete the overlying water of dissolved oxygen. Figure 2.13 shows the concentration of dissolved oxygen in water overlying enriched sediment (NCC, 1990). The site is a small Scottish freshwater loch, in which rainbow trout had been farmed in cages for just 18 months. Culture operations stopped in September, 1984.

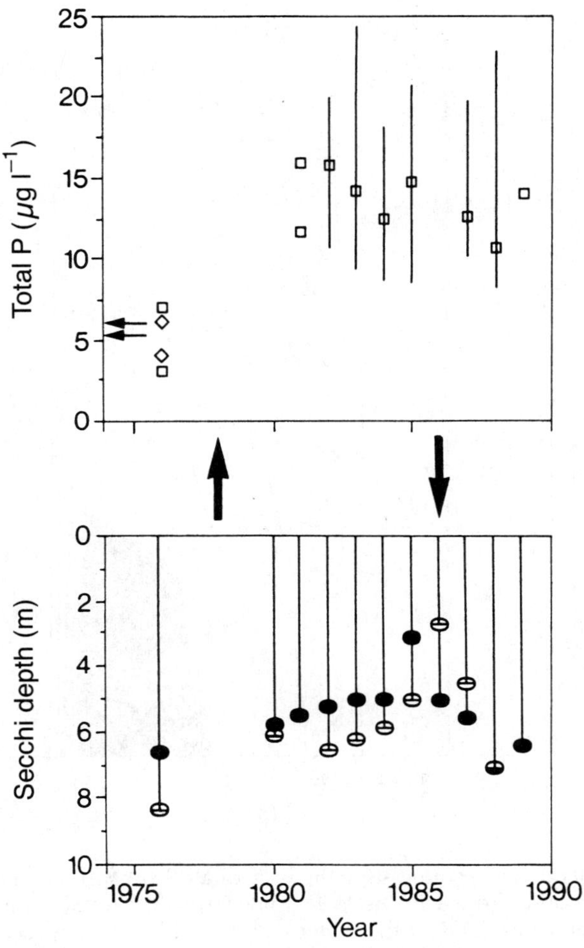

Figure 2.12 Potential effects of fish farming on Lake Skärsjön, Sweden: **(a)** total phosphorus concentrations (μg l^{-1}); **(b)** secchi depth readings (m). Bars indicate range of seasonal variation; □ phosphorus assessments during summers of respective years; ◇ samplings in 1972; ○ readings during June 1–15; ● readings during August 1–20; small arrows depict background concentration calculated by two methods; large arrows define period during which a fish farm was located in the lake. (From Persson, 1991, based on data from Enell, 1985.)

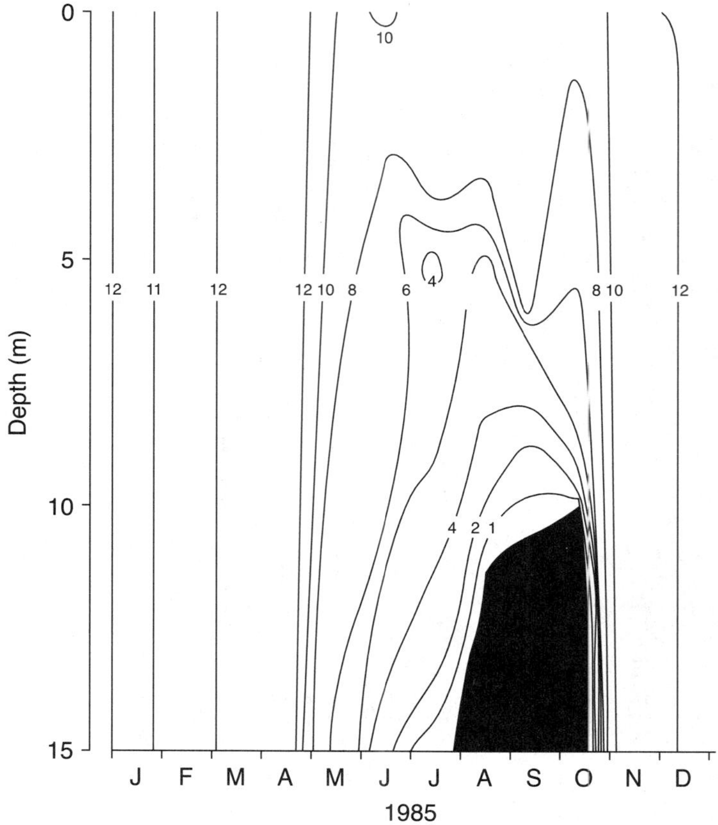

Figure 2.13 Dissolved oxygen isopleths in a small Scottish highland loch. Rainbow trout were farmed in the loch from April 1983 to September 1984. Shaded area indicates parts of the water column with a dissolved oxygen concentration of less than 0.5 mg l^{-1}. (From NCC, 1990.)

CASE 2.4: MUSSEL CULTURE (Mattsson and Linden, 1983)

The west coast of Sweden is an attractive environment for aquaculture, having many sheltered bays and channels. Although the focus is on salmon farming, mussel farming is also of considerable importance. It might seem at first sight that the environmental impact of mussel culture would be minimal, as no artificial feeds are supplied, in contrast to the situation for salmon farming. However, significant deposits of faecal solids and dead mussels can accumulate as a result of intensive mussel culture.

A study that investigated the effect of mussel culture on the benthic environment was undertaken off the island of Tjärnö, near Stromstad (Mattson and Linden, 1983). This region is characterized by an archi-

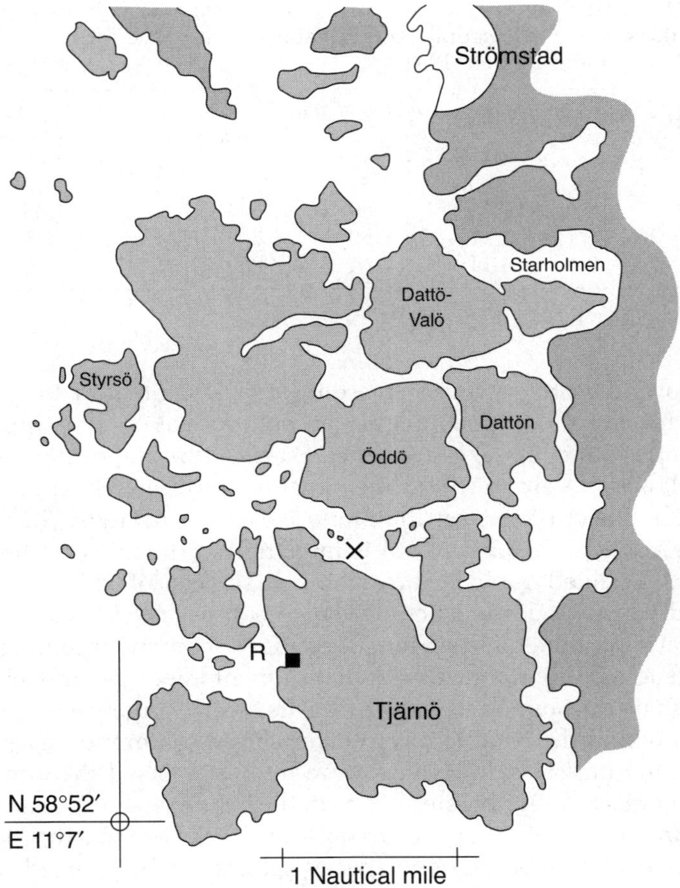

Figure 2.14 The Tjärnö region off the coast of Sweden: ×, location of mussel cultures; **R**, reference station. (From Dahlbäck and Gunnarsson, 1981.)

pelago of small islands, and it is in the lee of these islands that the study site for this investigation was located (Figure 2.14). The water in this sound varies from 8 to 13 m in depth. The tidal rise and fall is about 0.3 m, and currents are generally weak, 3 cm s^{-1} being a typical value.

Sampling sites were chosen beneath and in the vicinity of a block of longline mussel cultures (see Chapter 1 for details of this technique), and regular samples of benthic mud were taken over a 3-year period. The samples were analysed for their physical characteristics (particle size, organic content, etc.) and to determine the structure of their biological community.

Table 2.7 shows the physical characteristics of the sediment along a transect (a line of samples) from under the culture lines to 25 m distant.

Table 2.7 Grain size, organic content (ignition loss) and E_h in the uppermost 3 cm of the sediment, 15 months after establishment of the mussel culture (from Mattsson and Linden, 1983)

Sampling station	*Distance from culture (m)*	*Grain size Ø*	*% Sand*	*Ignition loss of dry weight (%)*	E_h *(mV) at 2 cm sediment depth*
2	0	5.85	34	22	–210
3	5	5.65	39	15	–205
4	15	4.81	50	8	–75
5	25	4.14	64	8	+20

This shows the progressive increase in organic content of the sediment and increasing redox potential as the culture station was approached (sampling station 2). This was the result of the deposition of faecal material, and closely mirrored the situation encountered under salmon cages. The effect of this on the fauna is shown in Figure 2.15. *Nucula nitidosa* is a small bivalve which is intolerant of organic enrichment. Its numbers declined rapidly during the first year of culture and was replaced by the opportunistic polychaete worm *Capitella capitata*. Whilst this showed considerable seasonal variations in abundance, it appeared to have successfully colonized the environment under the mussel culture. The changes in abundance shown by these two organisms reflect the changes within the benthic macro-invertebrate community as a whole.

When the mussel culture lines were first established the numerically dominant macro-invertebrate species in the benthos were *Nucula nitidosa*, *Echinocardium cordatum* (a sea urchin) and *Ophiura* spp. (brittle stars). Over the next 15 months these species gradually disappeared from the sampling station under the culture lines. By the end of the first 6 months all the brittle stars had gone. *Echinocardium* progressively declined in number until, after 11 months, none were left, and *Nucula* had similarly disappeared after 15 months. The community was progressively taken over by opportunistic polychaetes. After 6 months a dense population of *Capitella* was recorded, and in the latter part of the first year *Scolelepis fuliginosa* and *Microphthalmus sczelkowii* were recorded. At its peak this community of worms amounted to over 40 000 individuals m^{-2}.

At a separate sampling station the recovery of the benthos was studied following a 3½-year culture period. The sea bed at this site, 6 months after harvesting, consisted of mussel shells and black anoxic mud. The E_h value was –170 mV at 10 cm depth, and the fauna consisted of polychaete worms and amphipod crustaceans (*Corophium insidiosum* and *Microdeutepus gryllotalpa*). Eighteen months after harvesting the E_h value of the surface sediments had increased from –40 mV at 6 months to +100 mV (at 1 cm depth). The community was dominated by the same species but now included a greater range.

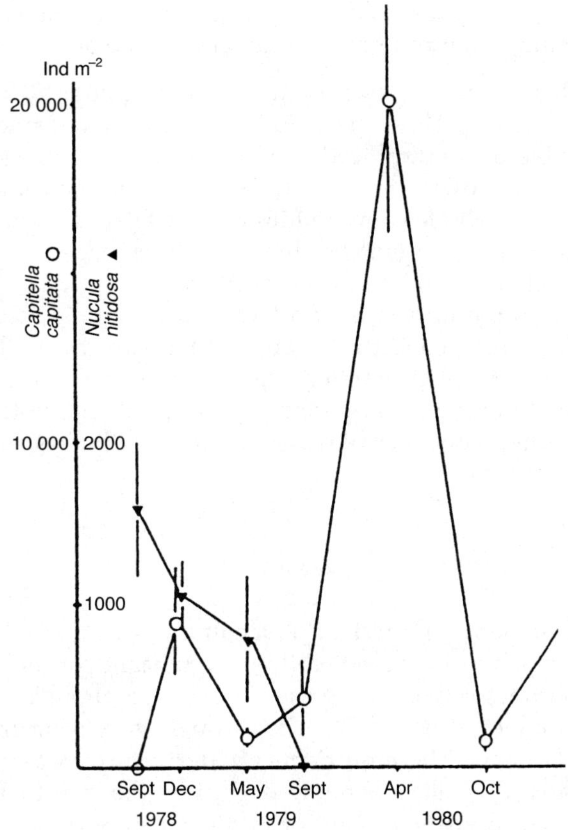

Figure 2.15 Number of *Nucula nitidosa* and *Capitella capitata* under mussel culture at sampling station 1, Tjärnö. After establishment of the culture (June 1978) the original macrofauna declined and was replaced by opportunistic polychaetes. (From Mattson and Linden, 1983.)

A final point of interest is the high number of fallen mussels recorded beneath the culture lines. Three months after the culture was established, 410 live mussels were recorded per square metre. The following spring, one year into the culture cycle, there were 2800 mussels m^{-2} with a biomass of 9.4 kg (wet weight with shells). These would all eventually die as they were smothered by falling mussels and faeces, and would make a significant contribution to the organic enrichment of the site.

2.4 CONCLUSIONS

The impact of organic wastes from aquaculture on the natural environment is mediated via:

- the release of nitrogen and phosphorus into the water column;
- the accummulation of organic sediments in the aquatic environment.

The result of these releases is to alter the ecological balance of the receiving ecosystem. Thus, the release of nitrogen and phosphorus into lakes can accelerate eutrophication and affect the dynamics and stability of the system. In flowing waters, the water quality and microbial, plant and benthic invertebrate communities may be altered. In marine systems the accumulation of organic sediments under cage farms can effect considerable changes to benthic communities.

In most cases the impact is of a localized nature, is recoverable, and is not of widespread concern. However, there are cases where aquaculture development is concentrated regionally, with resulting impacts on a wider scale. Intensive shrimp farming is of particular note, the regional impacts including poor coastal water quality and the destruction of large areas of mangroves.

SUMMARY

- The environmental impact of aquaculture is compared with that of other sectors, and is not generally of significant concern in the wider context (although there are some notable exceptions).
- The main effect of aquaculture is through increasing nutrient levels in natural waters. These nutrients are derived from particulate solids and dissolved wastes generated during the production process. Nitrogen and phosphorus are of particular importance.
- The consequences of the release of wastes from aquaculture into the environment include:
  - oxygen depletion;
  - eutrophication;
  - altered community structure.
- Case studies are presented to illustrate the nature of impacts encountered. They include:
  - intensive shrimp aquaculture;
  - freshwater trout aquaculture;
  - floating cage aquaculture;
  - marine bivalve aquaculture.
- Concerns over the concentrated development of aquaculture regionally are noted.

CHAPTER 3

Environmental impact of chemicals in aquaculture

3.1 CHEMICALS USED IN AQUACULTURE

The vast majority of the chemicals used by the aquaculture industry are for the treatment or prevention of disease, although cleaning agents are also used (Pillay, 1992; GESAMP, 1997). The most common products commonly fall into three classes (Alderman and Michel, 1992):

- **Topical disinfectants**, used in the treatment of topical parasites (e.g. some bacteria, protozoa and fungi). The group includes a wide and diverse range of compounds, such as malachite green, formalin, salt, copper sulphate, potassium permanganate and quaternary ammonium compounds.
- **Organo-phosphates**, primarily dichlorvos, for eliminating sea lice in marine salmon farming, and crustacean predators of fish fry in cyprinid nursery ponds.
- **Antimicrobials**, including a wide and diverse range of compounds, summarized in Table 3.1.

3.2 ADMINISTRATION OF TREATMENTS

The administration of therapeutants to fish or other aquatic animals is rather different from administration to terrestrial livestock. In the latter case it is possible to separate individual animals for treatment, or to treat a whole herd by giving the treatment to each animal in turn. In the case of farmed aquatic animals this is usually not possible. This creates problems in the administration of the treatment, for which there are environmental implications.

Firstly, the stock size is usually numbered in tens of thousands, rather than the tens or hundreds common to terrestrial animal groups. This precludes the separation of individuals for treatment not only because of the large numbers, but also because any attempt would cause considerable stress to the rest of the stock and perhaps exacerbate health

Table 3.1 Major antimicrobial drugs used in aquaculture (Alderman and Michel, 1992)

Drug	*Product*	*Route*	*Dose*	*Indication*
Antibiotics				
β Lactams	Ampicillin	oral	50–80 mg kg^{-1} 10 days	Gram-negative bacteria
	Amoxycillin	oral	50–80 mg kg^{-1} 10 days	Gram-negative bacteria
Aminoglycosides	Neomycin	oral	50–80 mg kg^{-1} 10 days	Gram-negative bacteria
	Kanamycin	bath	20 mg kg^{-1}	
Tetracyclines	Tetracycline	oral	50–80 mg kg^{-1} 10 days	Gram-negative bacteria
	Oxytetracycline	bath	20 mg kg^{-1}	
	Doxycycline			
Macrolides	Erythromycin	oral	50 mg kg^{-1} 10 days (bathe eggs in 2 mg l–1 1 hr)	Bacteria kidney
Non-classifiable	Chloramphenicol	oral	50–80 mg kg^{-1} 10 days	Gram-negative bacteria
		bath	20 mg kg^{-1}	
Synthetic antibacterial agents				
Sulphonamides	Sulphamethazine			
	Sulphadimethoxine	oral	200 mg kg^{-1} 10 days	Gram-negative bacteria
	Sulphaguanidine			
Potentiated sulphonamides	Trimethoprim + sulphadiazine	oral	50 mg kg^{-1} 10 days	Gram-negative bacteria
Nitrofurans	Furazolidone	oral	50–80 mg kg^{-1} 10 days	Gram-negative bacteria
	Furaltadone			
	Nifurpirinol	oral / bath	10–50 mg kg^{-1} 10 days	Gram-negative bacteria
Quinolones	Oxolinic acid	oral	12 mg kg^{-1} 10 days	Gram-negative bacteria
	Flumequine			

problems. Secondly, in the early stages of the culture cycle the small size of fish fry or larvae, for example, also precludes individual therapy.

Treatments are often administered to the whole stock in the holding unit, with the occasional exception of high value stock such as brood-stock. The three available methods are:

- oral administration (by inclusion of the chemical therapeutant in the feed);
- immersion of the stock in a solution of the chemical;
- mass injection.

3.2.1 Oral administration

This would seem to be the ideal route for the administration of many therapeutic compounds, especially those of high value where waste minimization is of primary concern. However, there are difficulties.

Some species, particularly shrimps, are messy eaters and a relatively high proportion of waste and leakage to the wider environment cannot be avoided. Once the therapeutant has been ingested, its efficacy cannot be guaranteed. Some therapeutants are not efficiently absorbed into the body via the intestinal tract. Oxytetracycline, for example, is thought to complex with Ca^{2+} and Mg^{2+} ions and consequently passes through the fish and into the environment still in an active form (Lunestad, 1992). Another factor with environmental implications is that fish and invertebrates often lose their appetite when stressed and this may be caused by disease or parasitic infestation. The result may be that a higher proportion of the delivered feed goes to waste, thus increasing the amount of the chemical entering the environment.

3.2.2 Bath treatments

Bath treatments are restricted to water-soluble compounds, and should be administered within a closed environment. The water supply to tanks should be stopped for the duration of the treatment, and floating cages should be enclosed in impermeable bags before treatment of the stock. Whilst this reduces the amount of chemical used compared with treatment in an open system (in which a higher dose is needed to compensate for leakage to the wider environment), some of the chemical will be discharged to the environment when normal operations are resumed.

3.2.3 Mass injection

This method is expensive, irrespective of the cost of the therapeutant, because of the labour or equipment needed to carry it out. It is only viable when the value of individual fish is high; for example, mass injection is used to vaccinate salmon against furunculosis, a virulent bacterial infection. It can be undertaken by hand or using one of the machines now available. The environmental impact of this technique is negligible.

3.3 ENVIRONMENTAL CONCERNS

The environmental concerns over the use of chemicals in the open aquatic environment relate to:

- the direct toxicity of the compounds;
- the development of resistance to compounds by pathogenic organisms;
- the prophylactic use of therapeutants;
- the length of time they remain active in the environment.

3.3.1 Direct toxicity

Dichlorvos and malachite green are good examples of widely used treatments that are believed to have significant environmental effects, and progressively greater restrictions are being placed on the use of each in the aquaculture industry. Dichlorvos is an organophosphate, a class of chemicals widely used for the control of crustacean parasites. It is effective against any crustacea, so may cause harm to marine crustacea such as (for example) crabs and barnacles in the vicinity of salmon farms as well as to the target species. The significance of the impact is difficult to quantify and remains a contentious issue.

Tributyl tin (TBT) is an anti-fouling compound, rather than a therapeutant, which was widely used in the treatment of fish cage nets (it was even more widely used to prevent encrusting growth of seaweeds and marine invertebrates on the hulls of boats). It readily leaches into the water and has been found to cause serious damage to marine invertebrate life. Consequently, its use is now regulated in many countries.

Nearly all therapeutants used in aquaculture have some toxic effect, as they are essentially biocidal compounds used to kill bacterial or viral invaders or invertebrate parasites. The obvious exceptions are vaccines, which enable fish to combat infections using their own immune defences.

3.3.2 Resistance

The development of resistance to antibiotics used in aquaculture is of particular concern, and there are now numerous documented examples (Alderman and Michel, 1992). The problem stems in part from the limited number of antibiotics available for use in aquaculture. In many countries farmers tend to be conservative in their approach, preferring to use a chemical which they know works rather than change to a new one; and in some countries use of antibiotics is not under legislative control. In the UK, as in many other countries, antimicrobial treatments are prescribed by veterinarians under government licence and many have restricted use (Richards, 1992; Hankin, 1992). As bacteria can rapidly develop resistance to antibiotic compounds, rotational use of antibiotics for the treatment of bacterial infections is an essential tool in maintaining the efficacy of these drugs (Alderman and Michel, 1992).

The environmental implications of the emergence of resistant strains of bacterial pathogens is twofold. Firstly, it makes infections more difficult to treat, thus endangering the viability of the farming process. Secondly, there are implications for human health. Waters receiving aquaculture discharges, and therefore containing antibiotic residues, may also receive domestic waste (Alderman and Michel, 1992). Thus, human pathogens are exposed to antibiotics used in aquaculture operations and which may also be used in human medicine – chloramphenicol is a notable example. It is possible that these pathogens may develop resistance to some of the important drugs used in treating human illness.

3.3.3 Prophylactic use

The prophylactic use of antibiotics, i.e. their regular use to prevent the outbreak of infection as opposed to their use as a response to infection, is a particularly important contributor to the development of resistance to antibiotics by bacteria. In many shrimp farming regions the prophylactic use of antibiotics to guard against possible infection by the bacteria *Vibrio* spp. (Alderman and Michel, 1992) has led to the development of resistant strains. As a consequence, some regions have been so heavily affected by *Vibrio* infection that hatchery operations have been abandoned. The result of prophylactic use is that much more antibiotic is used overall than when it is used only to treat infections as they occur.

Prophylactic use of chemicals for the treatment of external parasites does not present the same concerns with regard to developing resistance. In this case the concern is over the quantities of chemical entering the aquatic ecosystem, and the severity of its toxic effects on the natural community.

3.3.4 Persistence of antibiotic residues in the environment

Establishing the nature and severity of the effects of chemical usage by the aquaculture industry on the natural environment is a difficult task. In such instances it is often necessary to turn to laboratory studies to gain some insight into what may be happening in nature.

In this context an investigation was carried out to see what the effect of antibiotics commonly used to treat farmed fish might be on the microbial community of an organic sediment from under a marine floating cage fish farm (Samuelsen, 1992). The investigation consisted of a series of laboratory-based experiments, backed up by field surveys at a Norwegian salmon farm. Three antibiotics were tested: oxytetracycline, oxolinic acid and furazolidone. There were three themes to this work:

- the persistence of the antibiotics in the environment;
- the effect of their presence in the sediment on the microbial population;
- the emergence of resistance to their antibiotic activity within the microbial population of the sediment.

Table 3.2 shows the half-lives (Box 3.1) of these compounds in laboratory experiments and field surveys. The first notable point is that furazolidone has a very short half-life in the laboratory experiment, whereas oxytetracycline and oxolinic acid both had much longer persistence times (Figure 3.1). Furazolidone is metabolized by bacteria, producing the compound 3-(4-cyano-2-oxybutylideneamino)-2-oxazolidone (Figure 3.2). This compound is thought to be relatively inactive biologically. In addition the laboratory experiments reported in Table 3.2 used pure preparations of the antibiotics mixed into the sediment, whereas in real situations the antibiotic will be bound into the feed or faeces and the half-life is likely to be somewhat longer.

BOX 3.1
HALF-LIFE

The half-life is a measure of the length of time it takes for half the compound to degrade or otherwise disappear. Remember that at the measured time half the dose of the compound still exists in the sediment, and also that the rate of loss from the sediment may not be linear – it may take much longer for the second half of the dose to disappear.

Table 3.2 Half-lives of oxytetracycline, oxolinic acid and furazolidone in the sediment from laboratory studies and of oxytetracycline under three cages at salmon farms (from Samuelsen, 1992)

Compound	*Cage 1*	*Cage 2*	*Cage 3*	*Laboratory study*
Furazolidone	–	–	–	18 hours
Oxolinic acid	–	–	–	48 days
Oxytetracycline	125 days	142 days	89 days	55 days

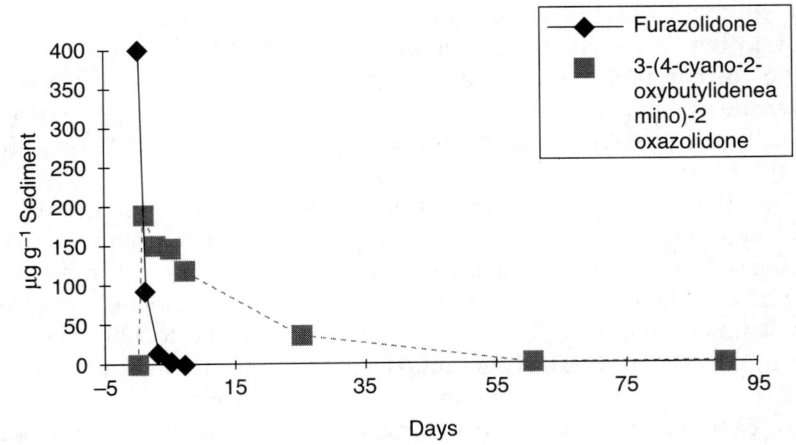

Figure 3.1 Metabolism of furazolidone and persistence of the main metabolite 3-(4-cyano-2-oxobutylideneamino)-2-oxazolidone in marine aquaculture sediment. (From Samuelsen, 1992.)

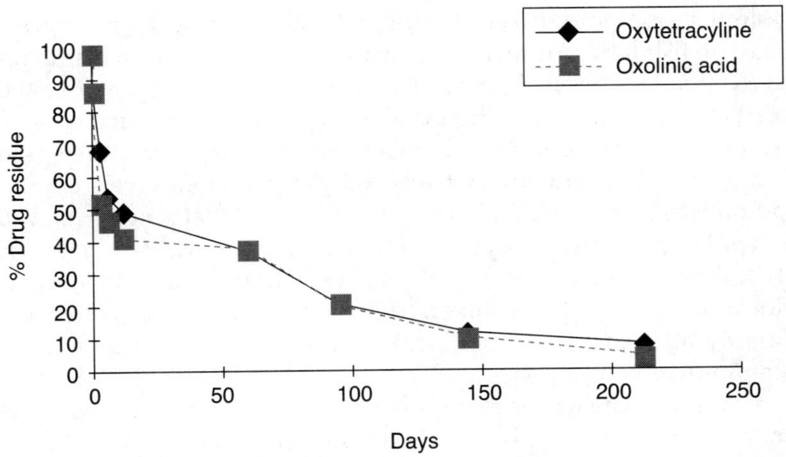

Figure 3.2 Persistence of oxytetracycline and oxolinic acid in a marine aquaculture sediment. (From Samuelsen, 1992.)

Oxytetracycline and oxolinic acid, on the other hand, are not known to be metabolized by bacteria (Samuelsen, 1992). Their loss from the sediment is probably by gradual diffusion into the overlying water since they are both water-soluble compounds. This assumption helps to explain the apparently anomalous measurements obtained in the field survey. Here, the half-life of oxytetracycline varied under different cages. The difference between cage 3 and the others is particularly marked. This may be explained by the observation that part way through the field survey (at the start of which oxytetracycline was administered orally to the fish in the cages) the fish were removed from cage 3, and so the further build-up of sediment under this cage was prevented. Under the other cages sediment continued to collect, covering the earlier deposits and possibly reducing the rate of diffusion of oxytetracycline into the water.

When this scenario was reproduced in the laboratory using crushed medicated pellets, instead of a pure preparation of the antibiotic mixed into the sediment (half-life 55 days; Table 3.2), the half-life was measured as 72 days. When a layer of medicated pellets was immediately covered with sediment, the half-life of oxytetracycline increased to 135 days. The clear implication of these observations is that the continued rain of sediment from the fish cages traps quantities of antibiotic in the sediment, where bacteria are exposed to it for much longer periods than would otherwise be the case (Samuelsen, 1992).

3.4 EFFECTS ON THE MICROBIAL COMMUNITY

Using similar laboratory-based experiments the effect of the presence of antibiotics in fish farm sediments on the microbial community was investigated (Kupka-Hansen *et al.*, 1992). Two days after treatment, with either oxolinic acid or flumequine, the number of bacteria measured in sediment samples had decreased by about a third. From this point they gradually recovered, but had still not reached pre-treatment levels 80 days after administration of the antibiotic (Table 3.3). Perhaps more relevant to the workings of the ecosystem are the measurements of bacterial activity following treatment (Table 3.4). Thus, the sulphate reduction rate (an indicator of the rate of mineralization of organic sediments) was significantly affected by the treatment, and took more than a month to recover pre-treatment levels.

The recovery of the bacterial populations is due, in part, to the development of resistance to the biological activity of the antibiotics. Bacteria are able to achieve this very rapidly due to their short generation time, and this characteristic is of significant environmental, public health and

Table 3.3 Total bacterial numbers in all sediments as a percentage of the control at different times after medication: numbers in brackets refer to total bacterial numbers (ml/g wet sediment) (from Kupka-Hansen *et al.*, 1992)

Treatment	*Day 2*	*Day 7*	*Day 24*	*Day 80*
Control	(3.0×10^{10})	(3.3×10^{10})	(4.3×10^{10})	(3.2×10^{10})
Control	100	100	100	100
Oxolinic acid	63	65	77	84
Flumequine	67	71	93	84

Table 3.4 Sulphate reduction rates (mM SO_4 m^{-2} per day) in all sediments at different times after medication (from Kupka-Hansen *et al.*, 1992)

Treatment	*Day 7*	*Day 29*	*Day 70*
Control	14	312	309
Oxytetracycline	0	55	412
Oxolinic acid	0	18	541
Flumequine	0	18	295

Table 3.5 Bacterial resistance towards the investigated antibacterial agents, in % plate counts, at different times after medication (from Kupka-Hansen *et al.*, 1992)

Treatment	*Day 2*	*Day 7*	*Day 24*	*Day 80*
Control	0.05	0.07	0.21	0.13
Oxytetracycline	2.8	13.5	–	8.7
Oxolinic acid	14.2	8.5	20.3	2.7
Flumequine	0.2	0.4	–	4.0

aquacultural concern. In the series of experiments cited, aerobic bacteria developed resistance to oxytetracycline and oxolinic acid after exposure for only 2 days. As the concentration of these antibiotics in the sediment declined, the bacterial resistance persisted at a high level (Table 3.5), though some reduction was observed at the end of the experiment (80 days after administration). Bacteria exposed to oxolinic acid also developed resistance to oxytetracycline and flumequine.

There are two points of interest in this work. The first is that bacterial populations develop a degree of resistance to antibiotics very rapidly, and are capable of relatively rapid recovery following treatment. In the absence of the drug this resistance is gradually lost from the population (over a period of weeks or months). Frequent or prophylactic use of the drug would maintain, if not increase, levels of resistance amongst the bacterial population. Secondly, the bacterial population in the sediments under the fish farm was severely affected with regard to its ability to metabolize the organic wastes derived from the fish farm.

Thus, frequent or prophylactic use of antibiotics would appear to have two potentially damaging consequences for the fish farmer:

- the development of resistant strains of pathogenic bacteria;
- a reduction in the rate of mineralization of the sediments under the fish farm, possibly leading to water quality problems at the site.

A study of bacterial resistance to antibiotics used in aquaculture was undertaken in Denmark. Here comparisons were made between bacterial populations from an unpolluted watercourse and from one with a high level of fish farming activity (Spanggaard *et al.*, 1993). The resistance level to oxolinic acid in the polluted watercourse was double that of bacteria in the unpolluted one. This clearly demonstrated the effect of antibiotic use in aquaculture on bacterial resistance to drugs in the wider environment.

A study in Ireland on the distribution of oxytetracycline residues under marine salmon cages indicated that the source of oxytetracycline in the sediments was probably waste feed rather than faecal waste (Coyne *et al.*, 1994; Kerry *et al.*, 1994). A predictive model was used to indicate the area of distribution of settled waste feed, and settled waste faecal material. The latter, having a lower density and settling rate, was dispersed more widely. Measurements of oxytetracycline residues were made in sediments that, according to the model, should contain only faecal waste, and sediments that should contain both feed waste and faecal waste. Oxytetracycline residues were not detected in the area subjected only to deposition of faecal waste.

3.5 IMPACT ON AQUACULTURE

In addition to effects on the wider environment, the development of bacterial resistance causes problems for the aquaculture industry (Michel and Alderman, 1992). A case of antibiotic resistance recorded in India highlights the problems that antibiotic resistance can cause. Very high mortalities were recorded in a shrimp hatchery of *Penaeus monodon*. Mysis, zoeal and post-larval stages were all seriously affected, with 70–90% mortality being observed (Karunasagar *et al.*, 1994).

Samples of larvae, of water within and outside the hatchery, and of the larval food sources were taken. The larvae had been affected with the bacterium *Vibrio harveyi*, causing 'luminous bacterial disease'. The source of this infection was thoroughly investigated. Water samples from the intake stream of seawater showed that whilst *V. harveyi* was present in low numbers in the seawater, the filtration system was capable of reducing their numbers sufficiently to prevent an outbreak of disease before the water arrived in the hatchery. Furthermore, broodstock held in the maturation tanks and the shrimp nauplii were not infected with the bacterium. Samples of the food sources for the larvae, the algal species *Chaetoceros* and brine shrimp, showed that they could not be the source of the outbreak.

An analysis of the antibiotic resistance of *V. harveyi* samples isolated from the hatchery and from coastal water provided the first clue to the problem. Those bacteria isolated within the hatchery showed much greater resistance to antibiotics than those isolated from the incoming seawater (Table 3.6), and showed greater virulence than control isolates

Table 3.6 Antibiotic sensitivity of some *Vibrio harveyi* isolates collected from different parts of the hatchery (modified from Karunasagar *et al.*, 1994)

Antibiotics	*Seawater*	*Seawater*	*Post larval tank 12*	*Post larval tank 8*	*Larval tank 8*
Co-trimoxazole	S	S	R	R	R
Erythromycin	R	R	R	R	R
Streptomycin	S	S	R	R	R
Oxytetracycline	S	S	S	S	S
Neomycin	S	S	S	S	S
Chloramphenicol	S	S	R	R	R
Gentamicin	S	S	S	S	S

R = resistant, S = sensitive

Table 3.7 Mortalities (number of larvae dead/number tested) of *Penaeus monodon* post-larvae in the presence of various bacterial isolates at different concentrations (quantified as number of bacteria ml^{-1} water) (from Karunasagar *et al.*, 1994)

	10^2	10^3	10^4	10^5	10^6	*Nil*
V. harveyi from larval tank	2/50*	16/50	35/50	50/50	50/50	1/50
V. harveyi from seawater	3/50	5/50	12/50	22/50	40/50	2/50
Control bacterial isolates:						
PL_1 *Pseudomonas*	2/50	3/50	8/50	10/50	20/50	2/50
PL_2 *Pseudomonas*	3/50	2/50	6/50	12/50	22/50	3/50
PL_3 *Vibrio* sp.	4/50	3/50	5/50	14/50	18/50	1/50

when tested on live larvae (Table 3.7). This further supported the view that the incoming seawater was not the source of the infection.

The conclusion drawn from this research was that an antibiotic-resistant strain of *V. harveyi* had developed and persisted in the larval rearing tanks in the hatchery, as a result of the prophylactic use of chloramphenicol and cotrimoxazole to control bacteria. The hatchery had been used continuously, without drying out and disinfecting the larval tanks. It is thought that the luminous bacteria survived on organic matter present in the larval tanks, and gradually developed antibiotic resistance through prolonged exposure. Tests showed that they were resistant to concentrations of 1000 mg chloramphenicol l^{-1} – it is a particularly strong antibiotic. The prophylactic use of this in the hatchery had become completely ineffective in protecting the stock.

The hatchery was shut down and the tanks dried and disinfected. It is now operating again successfully, but with very limited use of antibiotics.

3.6 IMPACT ON FISHERIES

Aquaculture in many parts of the world often coexists with capture fisheries, and it is theoretically possible that fish caught in the vicinity of aquaculture operations may have high levels of antibiotic in their tissues.

To investigate this issue, a study was undertaken in Norway of marine fish and invertebrates, caught in the vicinity of salmon farm cages (Lunestad, 1992). Tissue samples were taken from crabs, mussels and a wide variety of fish. The average concentration of oxolinic acid in the muscle of wild fish at the end of a course of treatment for the farmed fish was 3800 parts per billion (ppb). This is considerably higher than the tolerable limits for food for human consumption. The highest concentration found was 10 000 ppb in the muscle of coalfish. Twelve days later, oxolinic acid was not detectable in any of the wild fish caught. So, although it appears that oxolinic acid is metabolized by the fish quite quickly, their tissues could contain high concentrations of the antibiotic whilst it is being used in the farm, and for a short period after.

SUMMARY

- Chemicals used in aquaculture are mainly therapeutants for disease control.
- Therapeutants can be administered by means of oral presentation, immersion or injection.
- The main environmental concerns over the use of therapeutants in aquaculture result from:
  - the direct toxicity of the compounds;
  - the development of resistance to compounds by pathogenic organisms;
  - the prophylactic use of therapeutants;
  - the length of time they remain active in the environment.
- The main environmental concerns from the use of antibiotics in aquaculture are:
  - the possible development of resistance to antibiotics by human pathogens, such as *Vibrio* spp.;
  - the increased difficulty of treating diseases of aquaculture stock due to the development of resistant strains of pathogens.

CHAPTER 4

Waste minimization

The aim of this chapter is to set out a range of management measures to minimize waste and to detail methods for reducing the pollution loading from intensive aquaculture systems (and some other aquaculture operations) without recourse to any water treatment processes.

4.1 WASTE MINIMIZATION AND AQUACULTURE PLANNING

Farm design and location as well as management practices can have a significant impact on the production of polluting wastes. These factors should be considered at the conceptual and design stages as they will influence the final form of the farm. It helps to approach this task in a systematic way, for which the technique of environmental impact assessment is recommended.

4.1.1 Environmental impact assessment (EIA)

EIA is a procedure for assessing the potential environmental impacts of new developments, with a view to minimizing those impacts through appropriate design (World Bank, 1991). In some countries an EIA is a legal requirement for any new aquaculture development.

As discussed in Chapter 2, poor siting and design of intensive aquaculture units can have serious financial consequences for the business (e.g. self-pollution in floating cage aquaculture). There are also benefits to be gained from a good environmental performance by aquaculture businesses (Chapter 7). So, whilst it may seem to be an added expense and complication, it is recommend that EIA be integrated into the farm design process, and indeed into the business feasibility study and business plan (Nash, 1995). How EIA might fit in with these procedures is illustrated in Figure 4.1. An important point to note from this illustration is that EIA is used in direct support of each stage of the project design and its implementation. Thus, it will have a major influence on the feasibility study, where different operational procedures will be evaluated for their environmental impact and economic feasibility. During the detailed design of the project and during project construction and

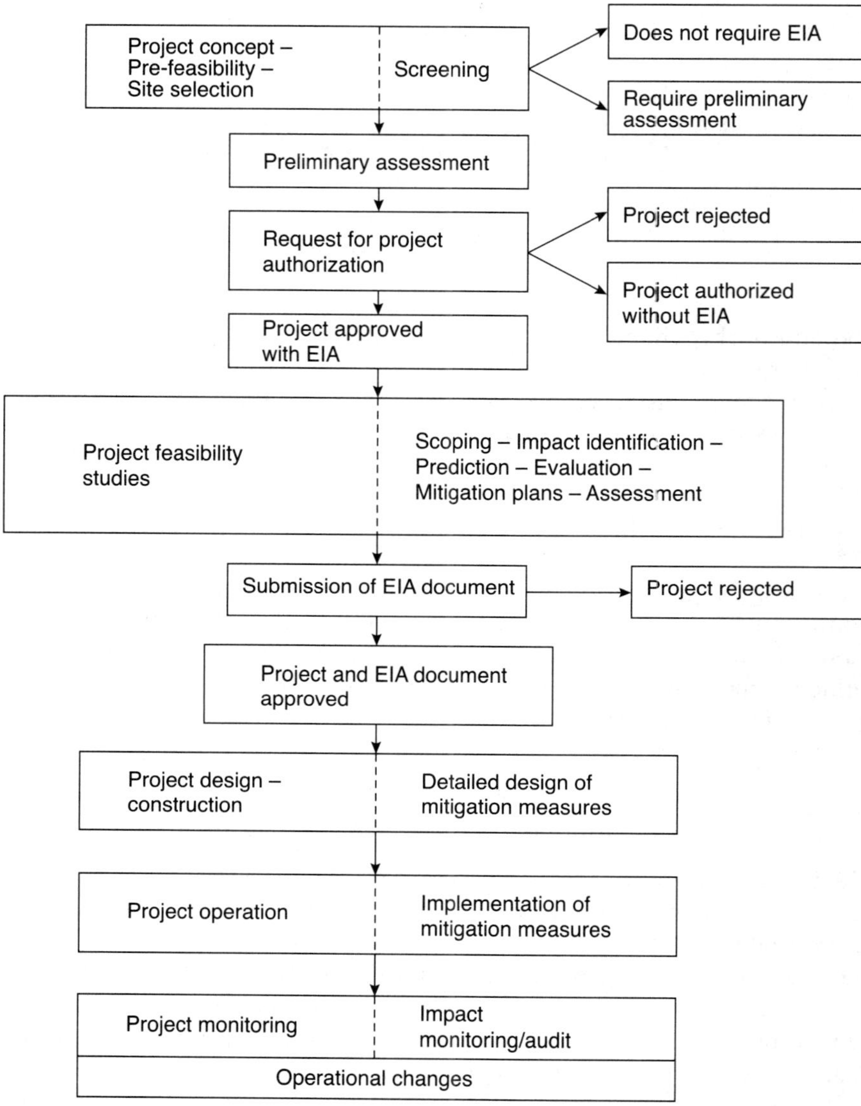

Figure 4.1 Simplified flow diagram showing integration of project activities with environmental impact assessment (EIA). (From Pillay, 1992.)

implementation, the EIA will provide an important reference work and a standard against which to set performance criteria (Pillay, 1992).

It is important, as a first step in the procedure, to define the scope of the EIA. A list should be compiled of all possible impacts from the proposed development, using past experience of similar aquaculture operations. The contents of this list should be assessed, using a scoring method where appropriate, so as to identify those impacts that are likely

Table 4.1 Defining the scope of an EIA for a new intensive fish farm

Category of impact	*Source of impact*	*Severity (score)*
Eutrophication of river	Nutrient discharge	
Oxygen depletion in river	Fish respiration/ effluent BOD	
Reduced flow in river	Abstraction	
Alteration of benthic substrate	Sedimentation of solids	
Prevention of migratory movements	Abstraction weir	
Destruction of wetland/mangrove (high wildlife value, user conflicts)	Pond/site construction	
Disturbance of wildlife	Construction activities	
Visual pollution (scenic area)	Predator nets/buildings	
Alteration of wild fish populations	Competition between escapees and wild fish, genetic interaction, pathogen transfer	
Hydrocarbon contamination of soil and water	Diesel oil spillage	
Alteration of microbial and invertebrate populations	Use of antibiotics and other chemicals	
Noise pollution	Blowers/pumps/vehicles	

to be of significant environmental concern. The EIA should be focused on these factors. Table 4.1 illustrates that adverse environmental impact is not confined to the aquatic environment, but in the case of aquaculture the most severe impacts are likely to be related to the release of pollutants into the aquatic environment as a result of the farming operations. Once the scope has been defined, baseline studies can be conducted to gather the range of information needed to make the assessment.

It should be noted here that EIA is not necessarily restricted to the natural environment. There may be impacts on the local social and economic environments as a result of the development. Furthermore, these impacts may not be negative. The development of the Scottish salmon farming industry, for example, has had many positive social and economic benefits for the remote communities of the Highlands and other areas of Scotland.

This exercise raises the question of how the potential severity of impacts can be judged in an objective way. There are a number of methods available for the quantitative assessment of impacts and one that is widely used is the Leopold matrix (Figure 4.2) (Pillay, 1992). It should be remembered that although this is a quantitative process, it is based on judgements that embody subjective elements and should always be interpreted with care. Some detail follows regarding the application of the Leopold matrix to EIA to illustrate the principles underlying analytical frameworks of this type.

Scoring system

Magnitude: 1 – small; 2 – medium; 3 – large

Importance: A – small; B – medium; C – large

* interaction; either no impact or positive impact

			Construction activities for preparing the site						Water and land regime alterations				Operational activities							Labour requirements			
			Canals, drains and ponds	De-watering	Construction traffic	Material stock piling	Structures	Vegetation clearance	Groundwater	Surface water	Flow regulation	Water characteristics and alterations	Pumping	Pond, drainage/filling	Water quality	Water management	Vegetation development	Manure storage	Offsite traffic	Construction workers	Permanent staff (technical)	Permanent staff (manual)	Permanent staff (guardians)
Physical and chemical characteristics	Earth	Land form	3/B			2/A																	
		Soil constitution	3/A	2/*	1/A	2/A			1/A	2/A													
		Mineral resources	1/A																				
		Construction materials	3/A																				
	Water	Surface	3/*	3/A		2/B	1/A			1/A				1/A	*								
		Groundwater	1/A	1/B		1/C				1/A	1/A		1/A	1/A	*			1/C					
		Recharge	1/A	1/A		1/A							1/A	1/A	*								
		Quality	1/A	3/*		1/C	1/A																
	Air	Climate								1/A			1/A	1/A									
		Quality				1/A	1/A											1/C					
	Processes	Floods	*	*																			
		Erosion	*	*				1/A															
		Silting		*																			

Figure 4.2 EIA matrix of a project for the establishment of a fish farm in the Nile delta in Egypt. (From Pillay, 1992.)

Cultural factors	Land and water use	Agriculture	1/A	*	1/B				1/A						*								
		Residential			1/A				1/A														
		Grazing	2/B		2/B	1/A		3/B							*								
		Gathering (fuel)	2/A		2/A	1/A		3/B	*														
		Gathering (wild crops)	2/A		2/A	1/A		2/A	*														
		Fishing																					
	Aesthetics and interests	Wilderness	3/B		3/A	2/A																	
		Historical sites	1/C	*	1/B	1/A				1/A						1/C							
		Recreation		1/A																			
	Human aspects	Cultural patterns																		1/C	1/C	1/C	
		Settlements				1/A			1/A				1/B		*			1/B					
		Employment	*	1/A				1/A												*		*	*
		Health aspects			1/A	1/C								*	1/C	1/C	1/C	1/C	0/C	1/A	1/A	1/A	
Biological status	Flora	Trees and shrubs	3/A					3/A	*														
		Grass and weeds	3/A			1/A		3/A	*								*						
		Aquatic plants	B/A	3/A				3/A	*								*						

Continued

Figure 4.2 Continued.

Scoring system

Magnitude: 1 – small; 2 – medium; 3 – large

Importance: A – small; B – medium; C – large

* interaction; either no impact or positive impact

			Construction activities for preparing the site						Water and land regime alterations				Operational activities							Labour requirements			
			Canals, drains and ponds	De-watering	Construction traffic	Material stock piling	Structures	Vegetation clearance	Groundwater	Surface water	Flow regulation	Water characteristics and alterations	Pumping	Pond, drainage/filling	Water quality	Water management	Vegetation development	Manure storage	Offsite traffic	Construction workers	Permanent staff (technical)	Permanent staff (manual)	Permanent staff (guardians)
Biological status	Fauna	Insects	3/A			*		1/A		*							*	2/B					
		Aquatic animals	B/B	3/C						*													
		Birds	*	1/A	1/A	*		1/A		*				*			*	*					
		Land animals	B/A	1/A	1/A	*		2/A									*						
Manmade works	Existing	Industries (salt extraction)		2/A					2/A	2/A	2/A		1/A	1/A		1/A							
		Structures	1/C		1/A																		
		Military	*		2/B														1/A				
	Proposed	New Ameriya City																			*	1/B	*
		Recreation																			*	*	
		Free trade zone														2/C							
		Others										1/A	2/B										
Disease incident		Schistosomiasis	1/A	*						1/C	1/C	1/B		*	1/C	1/C	1/C			1/C			
		Malaria	1/A	*						1/C	1/C	1/B		*	1/C	1/C	1/C			1/C			
		Others	1/A	*	1/A					1/C	1/C	1/B		*	1/C	1/C	1/C	1/C		1/C			

Figure 4.2 Continued.

The Leopold matrix allows the determination and quantification of the effect of individual operational procedures on individual components of the environment. These are selected by the assessor, according to their importance for each particular development. Each interaction is scored according to its **magnitude** (i.e. insignificant to severe) and its **importance**. For example, the noise level of construction machinery may be severe, but considered of low importance because of the remote location of the site. On the other hand, the eutrophication of a lake by a floating cage fish farm may be regarded as of little or intermediate significance in magnitude, but of high importance because of its wider environmental implications.

The individual scores for importance and magnitude can be multiplied together to give an impact factor (it may be either positive or negative) and the rows and columns of the matrix can each be summed to provide a relatively objective assessment of:

- the impact on different components of the ecosystem;
- the impact of different operational procedures on the ecosystem as a whole (this must, however, be interpreted with caution);
- a combined factor for the project as a whole.

4.1.2 Predictive modelling and EIA

Predictive models can be used where they are available to assist in the determination of the impacts of the development. In British Columbia, Canada, for example, the regulatory authorities use predictive models to assess the potential environmental impact of marine salmon farms when considering applications for fish farm licences (Caine and Castledine, 1991). The same models can be used to assist in redesigning components of the project where an unacceptable impact has been identified. This is an iterative process, in which different facility designs, site locations or operational procedures can be assessed with respect to their environmental performance.

Unfortunately few such models exist. Two that have been used in relation to salmonid farming are the Dillon and Rigler model of freshwater lake eutrophication (Beveridge, 1987) and a model of the dispersal of solid wastes from floating cage fish farms (Gowen *et al.*, 1989).

(a) Modelling lake eutrophication

In the Dillon and Rigler model the total phosphorus concentration [P] in a lake is a function of its volume (area × mean depth), its flushing rate (the proportion of its volume flowing out of the lake annually), the phosphorus loading on the lake, and the fraction of the phosphorus lost permanently to the sediments:

$$[P] = L(1 - R)/\bar{Z}\rho$$

where [P] is concentration of total-P ($g\ m^{-3}$), L is the total-P load ($g\ m^{-2}$ per annum), $\bar{Z}$ is the mean depth (m), R is the fraction of the total-P retained by the sediments, and ρ is the flushing rate in volumes per year.

Beveridge (1987) suggested the following three steps in determining the maximum level of aquaculture production, in relation to the maximum desirable [P] for a particular lake (this will vary with existing trophic status and with climate).

1. Measure the steady state total-P concentration. This is best done in the spring in temperate lakes when waters are well mixed. In tropical lakes the mean annual total-P concentration of surface waters will have to be determined by a long-term sampling programme.
2. The capacity of the water body for intensive cage fish culture is the difference, $\Delta[P]$, between the phosphorus concentration of the water body prior to exploitation, $[P]_i$, and the maximum desirable phosphorus concentration, $[P]_f$. This will vary depending on a variety of factors, the existing status and uses of the lake being the most important (Figure 4.3).

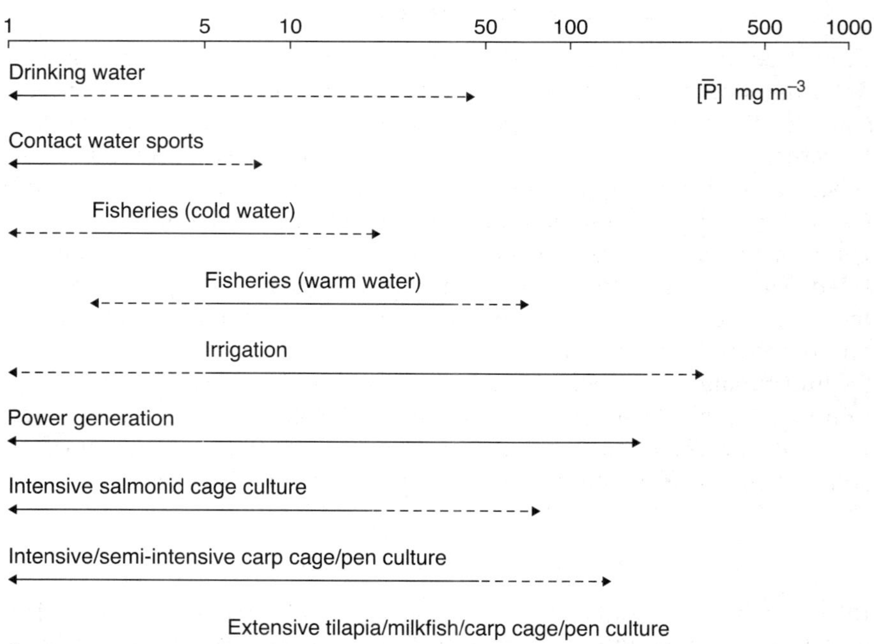

Figure 4.3 Suggested acceptable (dotted line) and ideal (solid line) P concentrations associated with freshwater bodies used for different purposes. (From Beveridge, 1987.)

Thus:

$$\Delta[P] = [P]_f - [P]_i$$

$\Delta[P]$ is related to phosphorus loadings from the fish culture operation, L_{fish}, the size of the lake, A, its flushing rate and the phosphorus adsorption capacity of the lake sediments:

$$\Delta[P] = L_{fish}(1 - R_{fish})/\bar{Z}\rho$$

$$L_{fish} = \Delta[P]\bar{Z}\rho/(1 - R_{fish})$$

where $\bar{Z}$ = volume V (m^3) divided by area A (m^2); ρ (y^{-1}) = average total outflow $(Q_o)/V$. $R_{fish} = x + [(1 - x)R]$, where x is the net proportion of total-P lost permanently to the sediments as a result of solids deposition (45–55% generally) and R is the proportion of dissolved total-P lost to the sediments, calculated according to Table 4.2.

3. Once the maximum acceptable phosphorus loading, L_{fish}, has been determined, then the intensive cage fish production ($t\,y^{-1}$) can be calculated by dividing L_{fish} by the average total-P wastes per tonne of fish produced. Box 4.1 gives a worked example.

Whilst the use of this model as a site selection aid will not by itself reduce the pollution load on the environment, it will help to ensure that a lake is chosen with the right characteristics to be able to deal with that waste without detriment to its normal functioning as an ecosystem, or to ensure that the farming operation is an appropriate size for the assimilative capacity of the lake.

(b) Marine impact model

This predictive model is detailed in Gowen *et al.* (1989) and will be referred to here as the marine impact model. It was designed to assist in the evaluation of sites for floating cage farming of salmon, though it is applicable to any type of floating cage aquaculture in both marine and freshwaters.

The marine impact model is illustrated in Figure 4.4. The current velocity and water depth obviously influence the horizontal distance travelled by a settling waste particle before it reaches the bottom. The size and density of the particle will also influence the dispersion distance, because heavier and larger particles sink faster. There is, therefore, less time for the horizontal force of the current to displace them. The critical variable is current velocity and direction, which will vary over a tidal cycle. Also, waste particles especially those of faecal origin, are unlikely to be uniform in size or density.

Water depth and current velocity and direction are directly measurable (the latter using, for example, hourly mean values of measurements with current recording meters). It is not possible to record continuously

Table 4.2 Empirical models for calculating the sedimentation rate σ, retention coefficient R ($1/\sigma$) and sedimentation coefficient V of phosphorus for general and specific categories of temperate water bodies (from Beveridge, 1987)

Model type	*Size of database*	*Model*	*Correlation coefficient*	*Source*
General	704	$\sigma = 0.129\ (L/\bar{Z})^{0.549}$	0.81	Canfield and Bachmann, 1981
US EPA		$R = 1/(1 + 0.614\rho^{0.491})$	*0.79	Larsen and Mercier, 1976
database		$\sigma = 0.94$	*0.79	Jones and Bachmann, 1976
and several		$V = 2.99 + 1.7q_s$	*0.73	Reckhow, 1979
European	73	$V = 5.3$	*0.71	Chapra, 1975
lakes and		$R = 1/(1 + \rho^{0.5})$	0.79	Larsen and Mercier, 1975
reservoirs		$\sigma = 0.65$	0.79	Jones and Bachmann, 1976
		$R = 0.426\ \exp(-0.271q_s) + 0.574\ \exp(-0.00949q_s)$	0.71	Kirchner and Dillon, 1975
		$V = 11.6 + 1.2q_s$	0.68	Reckhow, 1979
		$\sigma = 10/\bar{Z}$	0.68	Vollenweider, 1975
		$V = 12.4$	0.66	Chapra, 1975
Reservoirs	210	$\sigma = 0.114\ (L/\bar{Z})^{0.589}$	0.83	Canfield and Bachmann, 1981
North American		$R = 1/(1 + 0.515\ \rho^{0.551})$	*0.80	Larsen and Mercier, 1976
Natural lakes	151	$\sigma = 0.162\ (L/\bar{Z})^{0.458}$	0.83	Canfield and Bachmann, 1981
		$R = 1/(1 + 0.747\ \rho^{0.507})$	*0.80	Larsen and Mercier, 1976
Lakes with low flushing rates ($q_s < 10$ m)	53	$R = 0.201\ \exp(-0.0425q_s) + 0.574\ \exp(-0.00949q_s)$	–	Ostrofsky, 1978

q_s = areal water loading (mg^{-1})
ρ = flushing rate (volumes per year)
* = coefficients recalculated by Canfield and Bachmann (1981) using their database

BOX 4.1
ASSESSING LAKE CARRYING CAPACITY FOR CAGE AQUACULTURE OF RAINBOW TROUT

This example is taken from Beveridge (1987)

Site

Climate: temperate
Lake surface area, A: 100 ha (calculated from map)
Mean depth, $\bar{Z}$: 10 m (determined by hydrographical survey)
Flushing coefficient, ρ: 1 y^{-1}.

Method

Step 1: determine the steady-state [P], i.e. $[P]_i$, prior to development
15 mg m^{-3} determined by a monitoring programme for this site.

Step 2: set maximum acceptable [P], i.e. $[P]_f$, following the introduction of fish cages
Assuming no other developments or criteria take precendence, 60 mg m^{-3} is chosen as target $[P]_f$. Note: the target $[P]_f$ should be individually determined for each water body.

Step 3: determine ΔP

$$\Delta P = [P]_f - [P]_i = 45 \text{ mg m}^{-3}$$

Since $[P] = L_{fish}(1 - R_{fish})/\bar{Z}\rho$:

$$L_{fish} = [P]\ \bar{Z}\rho/(1 - R_{fish})$$

and:

$$R_{fish} = x + [(1 - x)R]$$

where R is calculated from Table 4.2, and x is assumed to be 0.5. Thus:

$$L_{fish} = 45 \times 10 \times 1/0.23 = 1957 \text{ mg m}^{-2}\text{ y}^{-1} = 1.957 \text{ g m}^{-2}\text{ y}^{-1}$$

Step 4
Since the lake has a surface area of 10 million m^2, the total acceptable loading = 1.957×10^6 g y^{-1}.

Therefore, the tonnage of fish that can be produced, assuming a P loading of 17.7 kg t^{-1}, calculated using the formula: weight of P in food fed × 0.48 (percentage assimilated by fish and not lost to the environment – this value is different for different species), is:

$$1.957 \times 10^6 \text{ g}/17\,700 \text{ g} = 111 \text{ t y}^{-1}.$$

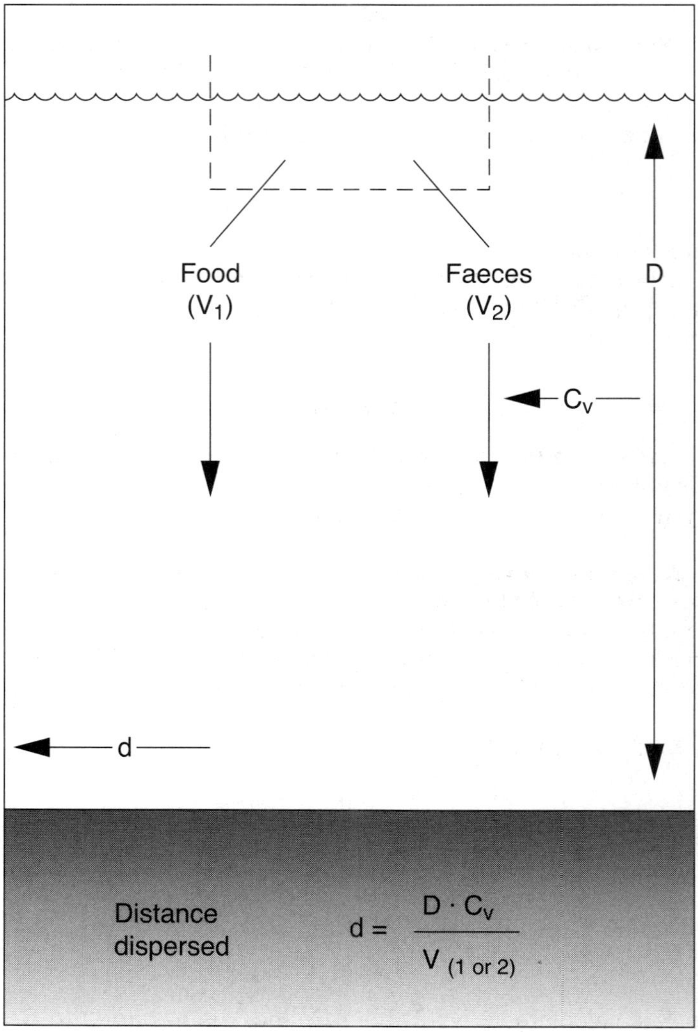

Figure 4.4 Horizontal displacement of organic waste: relationship between water depth, current velocity and settling velocity of waste particles. (Redrawn from Gowen *et al.*, 1989.)

the density and size of the settling particles to provide data to put into the model, and some assumptions have to be made. The originators of the model suggested the use of separate experimentally determined values for the two categories of waste: faecal solids and waste feed. In their validation of the model the authors used 0.12 m s^{-1} as the settling velocity of waste feed, and 0.04 m s^{-1} for faecal solids (this data is taken from Warrer-Hansen, 1982, and should be substituted by appropriate data to the culture system and type of feed used).

Up to this point the model is fairly simple. However, the objective of the model is to chart the dispersal pattern and eventual settling point of all individual particles leaving the cages. The cages cover a relatively large area, not just a single point, and so the model has to account for multiple points of origin of waste particles. Thus, the area occupied by the cages is divided into a grid of units 1 m^2, with coordinates *I* and *J*. Thus one might refer to one square by the coordinates 3*I*.4*J*.

Now it is possible to compute where a particular waste particle, leaving the cage at a particular stage in the tidal cycle and from a particular point, will settle (Box 4.2). Using this process a picture can be built up of the pattern of sedimentation from a cage farm in a particular location. Knowing the input of feed to the cage, and having estimates of resultant waste production, it is also possible to determine the total deposition of organic waste at each point on the sea bed. Figure 4.5 shows the predicted waste deposition pattern, using real data, from two Scottish floating cage salmon farms. At farm A the maximum recorded

BOX 4.2
MARINE IMPACT MODEL: CALCULATION OF DISPERSAL VECTORS

Using hourly measurements of current speed and direction the horizontal displacement of waste solids is given by the equations:

$$I + (D \times V\cos\theta)/U$$

$$J + (D \times V\sin\theta)/U$$

I and *J* are the coordinates of the waste particles within the grid at the start of each hour. The length of time the particle remains in the water column and therefore subject to horizontal displacement is given by *D* (water depth) divided by *U* (settling velocity of feed or faeces). $V\cos\theta$ and $V\sin\theta$ describe the horizontal displacement of a particle. Thus these equations give the *x* and *y* coordinates on the sea bed in relation to the position of the farm at the end of each hour.

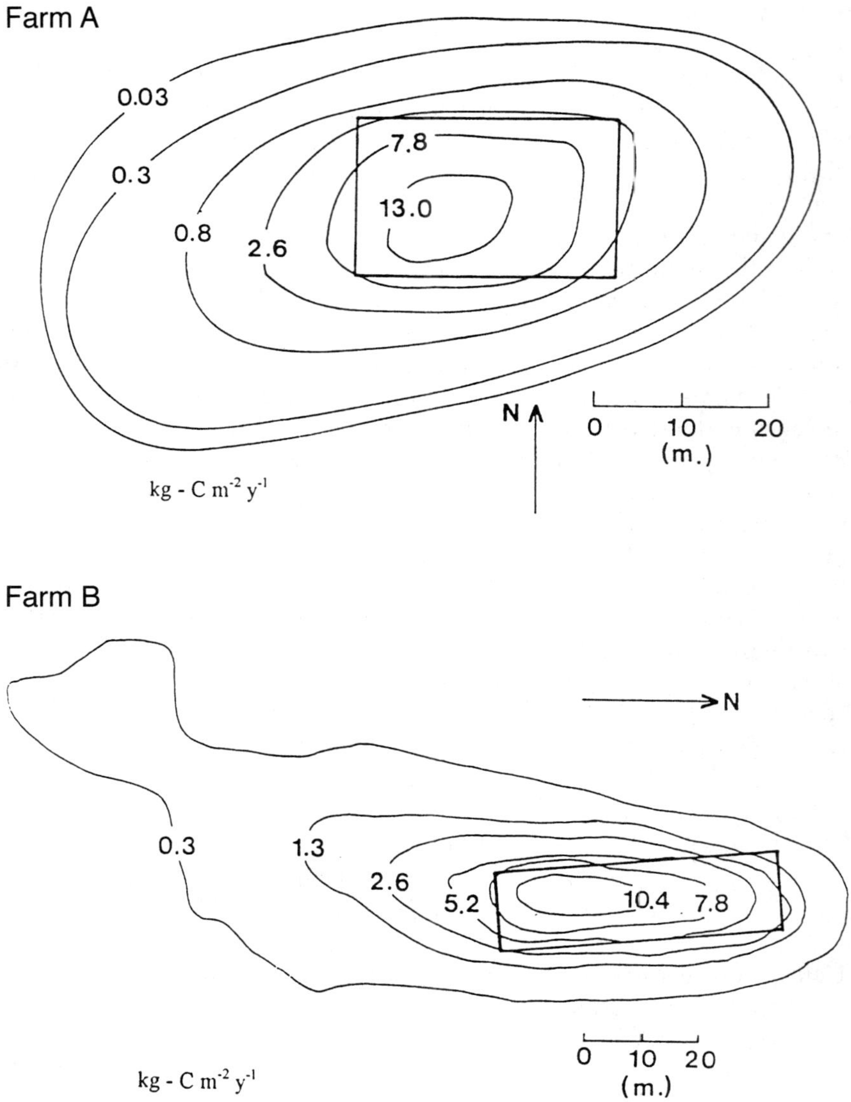

Figure 4.5 Marine Impact Model predictions for dispersion of waste solids from two floating cage farms in different conditions: see text for explanation. (From Gowen *et al.*, 1989.)

current speed was 0.16 m s^{-1} (very slow), with only a very weak south-westerly flow. By contrast, at Farm B the maximum recorded current speed was 0.42 m s^{-1}; there was a strong tidal component to the flow and a strong directional element. The influence of current strength and directionality can be clearly seen in the predicted dispersion patterns.

Table 4.3 A comparison between predicted loadings of organic carbon and measured sedimentary redox potentials at two Scottish salmon farms (from Gowen *et al.*, 1989)

Site	*Carbon loading* ($kg\ C\ m^{-2}\ y^{-1}$)	Eh_{4cm} (*mV*)
Farm A	7.8	–186
	5.2	–123
	2.6	–97
	0.8	–4
	0.3	41
	0.0	66
Farm B	7.8	–154
	3.3	–15
	0.2	12
	0.0	120

These predictions (calibrated in kg $C\,m^{-2}\,y^{-1}$ deposited) were compared with real data from the two sites. Here redox potentials had been measured to infer the carbon loading at particular points. In Table 4.3 the two sets of data, predicted and real, are compared. Although the values are not directly comparable, the trend indicates that the model had generated an acceptable approximation to the actual situation.

A second model proposed by the same authors, the hypernutrification model, did not accurately or even approximately reflect real nutrient or phytoplankton concentrations around the test sites. The dispersion of dissolved substances in a water body is subject to a whole range of forces, each often complex in nature, and which could not be incorporated into the model at the time of its conception. This failure illustrates the caution with which environmental models should be used. The more complex the environmental processes, the more likely it is that a model of them, which is of necessity highly simplified, will not be capable of providing a reliable picture of real situations.

Conversely, models that do achieve adequate representation of highly complex environmental processes are themselves complex. Whilst one or two models have been developed to represent the dispersal of dissolved inorganic nitrogen from marine floating cage fish farms and its impact on phytoplankton productivity (e.g. Silvert, 1992; Falconer and Hartnett, 1993), they are not easy to use. Despite this, they can still make a valuable contribution to the planning process (Silvert, 1992).

4.1.3 Other considerations

In lakes or semi-enclosed marine environments the hydrography is vitally important in site selection. Stratification in lakes or fjordic coastal embayments may, in association with cage aquaculture, cause deoxygenation and hypernutrification of the hypolimnion. This may have serious consequences for the aquaculture operation, especially in temperate lakes. In coastal regions, water currents – their direction, velocity and periodicity – are of more importance, though flushing rate is of significance in very sheltered fjordic embayments.

Trophic status of the environment is also important, especially in upland streams and oligotrophic lakes. It may also be important in semi-enclosed sea areas, where the trophic status may already be elevated through pollution from terrestrial activities, and may present the risk of toxic algal blooms or seasonal deterioration of water quality. The same is also true of many tropical lakes, especially if in a heavily populated area. However, most of the fish species farmed in tropical regions are less sensitive to eutrophic conditions than salmonids.

4.2 FARM DESIGN

The design of an aquaculture facility can have a significant effect on the impact it has on the environment. Some of the more crucial aspects are discussed in this section.

4.2.1 Tank or raceway design

The hydraulic properties of fish holding facilities can be used, by appropriate design, to enhance pollution control in intensive fish culture. Circular rearing tanks have been widely employed in intensive fish farming because of their self-cleaning properties. With appropriately designed tank outlet devices, a high percentage of the solids can be separated from the water stream (Cripps and Kelly, 1996). Whilst this is really a solids removal device, it is mentioned in this chapter because of the importance of specifying such devices at the design stage (see Chapter 5 for more details).

Raceways can also be designed with self-cleaning properties, by incorporating baffles along their length. At the end of the raceway a settling area can be constructed, using a different baffle design, where a high proportion of the solid waste can be collected (Wagner, 1993; Westers, 1991). This facilitates its removal from the system, thus reducing the pollution load on the environment.

4.2.2 De-intensification

Manipulating the degree of intensification of the production process should not be overlooked as a means of reducing pollution. In certain cases it can be economical, especially where self-pollution may occur. In the future, fiscal measures (e.g. taxes on pollution, subsidies for pollution control equipment; Nash, 1995) will favour less intensive aquaculture methods.

The environmental benefits of de-intensification are reduced nutrient and chemical residue loads. Stress levels in fish are related to stock density, which is why high stock density contributes to the incidence of disease. Reduced stock density should result in less disease and less chemical usage (O'Laoire and Welford, 1994).

Studies on the economics of intensification in shrimp farming in Thailand have shown that semi-intensive farming is more profitable than intensive production methods (Csavas, 1990). This is manifest in two aspects:

- price elasticity;
- long-term investment return in relation to sustainability.

Intensive shrimp farming requires more capital investment, in machinery such as aerators and water pumps, and has proportionately higher running costs (e.g. power supply for aeration and water pumping, and pond maintenance). Consequently, profit margins are low and farms are highly reliant on product turnover for continued financial viability – hence the pressure to maintain high levels of output.

Semi-intensive shrimp farms tend to have higher profit margins than intensive farms, as a result of lower running and financing costs per unit of production. Financial analyses have shown that whilst market conditions are stable, extensive, semi-intensive and intensive shrimp farming is profitable, but that only semi-intensive shrimp farming can accommodate a 20% fluctuation in input costs or market price (Primavera, 1991).

Experience in other aquaculture sectors has shown that initially profitable industries encourage overinvestment. The consequences are overproduction, market saturation and falling market prices. The shrimp farming industry has already suffered many failures for this reason, especially among intensive producers. An additional problem is that of environmental pollution, another consequence of intensive production methods and overinvestment. Long-term sustainability is threatened by intensive farming methods which are both too expensive and environmentally damaging (Briggs, 1994).

4.3 OPERATIONAL MANAGEMENT FOR WASTE REDUCTION

Good farm design is wasted without good husbandry practices. The use of EIA can identify the best environmental option for design and location, and in the same way the best husbandry practices can be identified with regard to waste minimization. The most important of these are discussed in the following paragraphs.

4.3.1 Cage site rotation

The reduction of the effects of solid wastes from cage farms is particularly difficult. Effective feeding strategies are important, but cannot solve the problem completely. There will still be a gradual build-up of organic sediment below the cages leading to benthic community changes, unless the site is very well flushed.

Site rotation has become an accepted method for reducing the self-polluting effects of organic enrichment of the benthos under floating cage farms. The length of fallow period needed to ensure adequate recovery (i.e. mineralization of a significant proportion of the organic matter in the sediment) will depend on the characteristics of each site, but an absolute minimum of one year is to be expected (Ley, 1993). It is possible that the fallow period would be shorter if farming were stopped at a site before the sediment under the cage became azoic (with respect to the macro-benthos). The presence of oligochaete worms and other decomposers should enhance the rate of mineralization but little information is available on this theory.

Whilst site rotation has benefits for the fish farmer, and for the environment at a local level, it has been suggested that it may be detrimental on a wider scale, as site rotation only spreads the effects of nutrient enrichment over a wider area. Although rotation does not solve the underlying problem of waste production, it is nevertheless useful in controlling the environmental effects of pollution. From an ecological point of view, ecosystems can cope with nutrient enrichment to a certain extent, by means of the natural processing of organic waste by the decomposers in the food web. Once this system has been overloaded, the normal processes break down and recovery takes much longer. Therefore it is better to prevent overload, by spreading the nutrient load over a wider area, than to suffer ecosystem breakdown, even on a localized scale.

4.3.2 Management of abnormal solid loads

One of the factors influencing temporal variations in wastewater quality from intensive fish or shrimp ponds, tanks and raceways is the cleaning or flushing out of the facilities, to remove settled solids and algal growth, or for harvesting (Butz, 1992; Funge-Smith and Briggs, 1994; Schwartz and Boyd, 1995). The effect of these activities on wastewater quality is illustrated in Table 4.4, although each case will differ substantially. The difference between trout pond discharges and those of shrimp ponds is notable. In the latter, most of the sediment on the pond bottom is inorganic, derived from erosion of the pond banks. In the trout ponds it is mostly organic solids which settle to the bottom and are flushed out during cleaning operations. This is reflected in the values of BOD_5 and

Table 4.4 Comparison of water quality (mg l^{-1}) during normal operation, and tank/pond cleaning or draining for harvesting

Water quality	*Intensive trout**		*Intensive shrimp***		*Extensive carp**
	Normal	*Cleaning*	*Normal*	*Harvesting*	*Harvesting*
Suspended solids	1.5–11.4	17–8010	133 ± 89	429 ± 320	690–13050
BOD_5	1.9–4.4	40–1150	1.9 ± 1.5	14 ± 7	16–84
COD	8–24	50–2770	–	–	180–2270
Total P	0.03–0.13	0.14–3.1	0.14 ± 0.14	0.67 ± 0.21	1.0–11.2
NH_4-N	0.09–0.52	–	0.06 ± 0.06	3.65 ± 2.75	–
NO_2-N	< 0.29	–	0.02 ± 0.03	0.10 ± 0.01	13–34N
NO_3-N	0.47–3.58	–	0.11 ± 0.17	0.17 ± 0.10	–

Source of data: *Butz (1992); **Funge-Smith and Briggs (1994)

total-P, which are much higher during cleaning (relative to the normal values) for the trout farm effluents.

The significant pollutant loading on the receiving waters resulting from cleaning or harvesting activities could easily be avoided if the wastewater from these operations could be discharged into a settlement pond. Ideally this pond would be separate from other settlement facilities, so that the wastewater could be treated in batches. This would allow a long residence time and could improve the quality of the effluent discharged. The optimal residence time would be site specific, but in most situations an allowance of one, two or more days' residence time would be beneficial. The settling pond need only be a fraction of the size of an individual holding unit, unless cleaning operations are frequent or the number of holding units large (see Chapter 5 for more on settling ponds).

In earthen ponds used for static-water culture, the water quality at harvest is affected by seining prior to drainage. If seining is undertaken, the sediment stirred up during this activity should be allowed to settle for a few hours before the pond is drained (Boyd, 1995). During drainage, most of the solid load in the effluent is contained in the last 20% of the water, and this could be diverted to a suitable settling facility.

4.3.3 Reducing the use of chemicals

Pathogens, when present in fish stocks at harmful levels, are normally controlled by the application of biocides. Biocides are toxic chemicals that have a usually lethal effect on the pathogen at which they are targeted. The problem is that aquaculture biocides are normally water soluble and may enter the receiving waters, where they can do as much harm to natural crustacea, insect or microbial populations as they do to the fish pathogens. Methods that minimize the use of biocides in aquaculture include:

- **The use of vaccines against bacterial and viral infections**. Great advances have been made in the development of vaccines in the last decade, and have made a valuable contribution to fish health and environmental health. The development of machines for mass injection of fish has potential environmental benefits. Treatment by oral administration (by inclusion of the drug in the fish feed) may not be an efficient method. Individual injection by ensuring the administration of an accurate dose to each fish, without waste, would result in a significant reduction in the quantity of therapeutants used.
- **Studies on the ecology of pathogens**. This approach has led to the development of a new method for sea lice control, for example. Sea lice are a considerable problem in salmon aquaculture: heavy infestation causes lesions on the fish which often develop secondary infections. The widespread treatment of this parasite with organophosphate biocides (e.g. dichlorvos) has caused much concern, and studies have demonstrated damage to other crustacean species. Considerable efforts have been made to develop alternative treatments that cause less environmental damage. One treatment method is based on sea lice behaviour: it takes advantage of the phototactic response of sea lice and attracts them to a stroboscopic light source within the cage. This is designed to resemble the shimmering of salmon scales, which is thought to attract sea lice to the fish. Once concentrated around the light they are pumped out of the cage, collected and destroyed.
- **The use of cleaner fish to consume sea lice**. This is another example of biological rather than chemical control (Anon., 1994; Bergheim and Asgard, 1996). In salmon farming systems in Scotland, for example, goldsinny wrasse are introduced into the cages. They are effective predators of sea lice, and help to keep populations below a critical level. Thus less reliance is placed on the use of organophosphates for sea lice contol. With the provision of sufficient resources many more biological control solutions could be developed for the aquaculture industry.
- **Strategic approaches to pathogen control**. The essence of this concept is to find a balance between an acceptable level of pathogen control and an environmentally acceptable level of chemical use. This idea is explained more fully in the following paragraphs.

With regard to reducing the use of therapeutic chemicals in aquaculture there may be an advantage in changing the emphasis from one of minimizing mortality to that of achieving a predetermined level of survivorship (O'Laoire and Welford, 1994). For example, in a carp hatchery malachite green is used to prevent the growth of *Saprolegnia* fungus in batches of incubating eggs. If malachite is not used, a batch of eggs may occasionally be lost. In the approach suggested here, this loss is acceptable and is expected in the pre-production planning. To allow for this loss a few more female broodfish and incubation jars may be specified. The output of the system is therefore maintained while the use of chemicals is reduced.

In most cases this may not prove economic at present – certainly not in grow-out operations. However, if restrictions on the use of antibiotics, for example, were to be tightened then this shift in managerial emphasis might be essential.

Willoughby and Roberts (1992) investigated the strategic use of fungicides in salmon hatcheries to control the parasitic fungus *Saprolegnia parasitica*. They showed that the use of malachite green could be reduced from continuous addition (to achieve a concentration of 0.25 mg malachite green oxalate l^{-1}) to one 15-minute dose of 0.25 mg l^{-1} every 24 hours. Whilst this did not totally prevent fungal growth, it did strongly inhibit it. The authors suggest that this inhibition, combined with the fishes' own immune system, might be sufficient to prevent serious infection.

4.4 REDUCING THE ENVIRONMENTAL IMPACT OF CHEMICAL RESIDUES THROUGH OPERATIONAL MANAGEMENT

Once a chemical has been administered to the fish stock, at least a proportion will find its way into the environment. As noted in Chapter 3, the greatest concern lies with the use of antibiotics, but other individual chemicals, such as dichlorvos, also have worrying impacts on the environment.

Minimizing the impact could be achieved in several ways:

- The impact of antibiotics on the natural microbiota is related to its residence time in their habitat. The use of antibiotics that degrade rapidly or that are readily metabolized by micro-organisms would represent an improvement in the situation. On this criterion the use of oxytetracycline and oxolinic acid would be discouraged, and the use of furazolidone would cause less concern (Chapter 3). There are other factors that must be taken into account in prescribing a therapeutant, but before any decisions are made it would be necessary to have environmentally relevant information such as its half-life and its effect on higher forms of life.
- Where possible, the direct injection of antibiotics should be encouraged in order to minimize the waste inherent in oral administration.
- For administration of antibiotics during the early life stages of cultured animals, the use of biocarriers should be investigated. The successful transfer of antibiotic to larval fish has been accomplished using *Artemia* as a carrier, enriched with antibiotic through immersion in a solution of the therapeutant.

- With regard to the development of antibiotic resistance amongst bacteria, and especially pathogenic bacteria, a number of recommendations can be made:
  - Prophylactic use of antibiotics should be discouraged.
  - A range of antibiotics should be administered, alternating the use of each. Microbial exposure to each drug will thus be minimized. Reduction of the range of antibiotics available for use in aquaculture may accelerate the emergence of resistant strains.
  - It should be ensured that the sediment from farms, if it cannot first be removed from the water column, is dispersed as widely as possible. This will ensure that the concentration of the antibiotic in the environment is minimized. If resistance develops within the microbial community, it is less likely to operate at the dose rates used for treatment purposes.

A recent report for the United Nations Food and Agriculture Organizations provides further guidance on the safe and effective use of chemicals in coastal aquaculture (GESAMP, 1997).

4.5 FEED STORAGE AND HANDLING

The efficiency of feed utilization by the fish can be affected by poor handling and storage. Rough handling can cause physical damage to the feed pellets and will increase the percentage of dust in each batch of feed. Fine particles of feed are wasted, as the fish preferentially take the large pellets, and the dust is quickly swept out of the pond, tank or cage.

If feed is stored for too long a period before use, or under poor conditions (high humidity or excessive heat, for example), its nutritional quality can deteriorate, resulting in a greater proportion being voided as faeces. Conditions of high humidity can encourage fungal growth and loss of quality. In addition food palatability can be affected, reducing feed intake and increasing the proportion of waste feed. Pelleted feeds should be used as soon as possible after manufacture or by the shelf-life recommended, because vitamins degrade with time and lipids are liable to oxidation, which affects their nutritional value and palatability.

4.5.1 Feeding strategies

The development of 'high energy' or 'low pollution' diets (Box 4.3) has had a considerable impact on fish farm wastewater quality. The success of these diets is not only due to formulation strategies but also depends on good fish husbandry. In addition to feed quality, therefore, water quality and feeding method are vital factors in achieving reduced pollutant load.

Diet management is a fundamentally different approach to many of the other pollution control methods described in this book, which relate

BOX 4.3
NOMENCLATURE FOR MODERN PELLETED FEEDS

'**High energy**' or '**nutrient-dense**' diets typically have high levels of lipid and lowered protein content. They are formulated on the principle of protein sparing. This allows a lower protein content in the feed, without affecting fish growth. This in turn results in a reduction of the ammonia-N loading on the environment that is an end-product of protein metabolism.

'**Low pollution**' diet is a more general term, relating to a number of properties of the diet. Typically such a diet will incorporate protein sparing and will have a high digestibility value (therefore less is voided as faeces). It may also incorporate low phosphorus fish meal, thus reducing the P loading on the environment. This is of benefit to freshwater ecosystems, but it has been suggested that the use of low phosphorus diets in marine locations may upset the natural N : P ratio, a factor that is important in phytoplankton growth. This might lead to increased incidence of phytoplankton blooms in the vicinity of marine floating cage sites.

to the removal of pollutants, in either soluble or solid form. This section focuses on reducing the production of polluting wastes, which is a more environmentally sound approach as it increases the efficiency (in resource terms) of the production process. That is, a greater weight of fish is produced for less input of feed. This is a step towards greater sustainability in the aquaculture industry.

The development of low pollution diets is closely linked with stricter environmental protection legislation. Denmark has one of the largest trout farming industries in Europe and has led the way with environmental legislation to protect the environment from the adverse effects of intensive fish farming (Chapter 9).

It is interesting at this point to examine the history of feeds for intensive fish farming, which shows how, in the past, the evolution of feeds has been governed by economic and practical factors, rather than by legislation. The use of low pollution diets has been enforced by government legislation, though these diets were developed by the industry in response to a perceived benefit to the fish farmer, before the legislation arose. Whilst the severity of certain aspects of current Danish legislation is often discussed by the industry in Europe, it is evident that low pollution diets are an established part of intensive aquaculture with or without the legislation.

Kiaerskou (1991) describes the development of fish feeds in the context of the Danish trout farming industry. In the latter part of the nineteenth century the first intensive trout farms appeared in Denmark, and elsewhere in Europe. The fish were reared on a combination of natural food and slaughterhouse wastes. The industry in Denmark expanded rapidly and by 1913 there were 138 farms. During this period farmers started to use waste fish ('trash fish') and fish processing wastes as a source of feed for the trout. This was readily available in the fishing ports. The use of trash fish increased in the interwar years, but the persistent problem of freshness led to the development of alternative feeds (there was no refrigerated transport or chilled distribution chains to enable fish to be delivered to the farm before decomposition started). In the 1950s a joint effort between the Danish and the American trout farming industries resulted in the first dry feed for trout, which became commercially available in the late 1960s. The advantages of a dry feed over trash fish or offal include reduced transport costs, increased storage time, improved water quality and low risk of disease transference.

It was not until 1985 that the use of trash fish in Danish fish farms was finally banned, on environmental grounds (Kiaerskou, 1991). By this time the next major development in the fish feed industry was already under way, with feed companies working on the development of low pollution diets. This was in response to the recognition of the importance of water quality for optimal fish growth, the effect of feed and waste feed on this, and to the increasing demands from the environmental lobby for fish farmers to 'clean up their act'. Low pollution diets became commercially available during the mid 1980s. Once it had been shown how effective these diets were in reducing pollutant loading from fish farms, the legislation was developed to make their use compulsory. In 1989 the Danish government introduced legislation that compelled fish farmers to use low pollution diets and to feed optimally, i.e. to keep the food conversion ratio (FCR, the ratio of weight of food fed to the growth, in weight, of the fish) below a certain level.

The main principles behind the development of low pollution diets are:

- high digestibility of the raw material to minimize the amount of faeces;
- high concentration of energy and nutrients in the feed to obtain the lowest volume of feed used per day;
- optimal nutrient balance to give the best possible feed utilization and growth, and the use of materials such as low temperature fishmeal;
- food acceptability (taste, texture, attractants) to maximize feed intake and minimize waste;
- the use of more efficient binders and pelleting techniques to reduce leaching and increase pellet stability;
- a greater range of pellet sizes to match the gape of the fish.

These factors all relate to the 'design', or formulation, of the feed. To these can be added equally important factors related to fish husbandry:

- feeding method (demand, auto, hand) and feeding frequency;
- water quality, especially dissolved oxygen concentration.

4.5.2 Fish nutrition

It is also important to examine the basics of fish nutrition in order to understand the implications of these factors for pollution control. Any feed, be it a commercially produced pellet or a live animal or plant, has five basic nutritional components: proteins, lipids, carbohydrates, vitamins and minerals.

Both the types of these nutrients (quality) and the relative proportions of them in the feeds (quantity) are important in determining fish growth, i.e. in optimizing the fish's utilization of the food provided. What the fish needs from its food intake is the right amount of the correct amino acids (the 'essential' amino acids) and lipids (essential fatty acids), enough energy to cover its maintenance and growth requirements, and vitamins and minerals which are essential to facilitate many of the biochemical reactions which govern growth processes.

(a) Protein

Excluding water, proteins are the main constituents of the animal body. Animals therefore need protein for growth. Proteins are made from amino acids, sometimes called the building blocks of life. Some amino acids cannot be synthesized within the body of animals, and are known as the essential amino acids (Jobling, 1993). They are passed through the food chain from plant organisms, where they are synthesized during the process of photosynthesis. Animals must therefore consume plants or other animals to obtain the essential amino acids. Once these have been ingested they can be broken down into the constituent amino acids and reformed into the specific proteins needed by the animal for growth (a process known as anabolic metabolism). Proteins can also be broken down, or catabolized, to produce energy (catabolic metabolism). Energy is needed to sustain the biochemical reactions that are the basis of respiration and growth.

Artificial fish feeds must supply the necessary protein that the fish would obtain from eating animals or plants in the wild. The quality of the protein varies, depending on the source. Protein quality in fish feeds reflects the suitability of the particular proteins for the fish's metabolic needs. Generally, fish meal has proteins that are common to most fish species. These can be used directly or with little modification by the fish, which therefore uses less energy in assimilating them into tissue growth.

Proteins from non-fish sources are less suitable and may have to be broken down and reformed. They may also be deficient in one or more essential amino acids. Because the constituent proteins of the body tissue vary slightly between species, each species requires different amounts of each amino acid. When formulating diets it is important to match the amino acid profile of the diet as closely as possible to the amino acid requirement of the species for which the diet is intended. This can be done by careful selection of the ingredients, which usually contain a mixture of plant and animal proteins. For carnivorous fish, it is important to include a relatively high proportion of fish meal, so as to achieve a balanced amino acid profile.

Any protein that cannot be used by the fish for growth, because it is unsuitable (e.g. indigestible) or because other essential compounds are not present (see below), is catabolized for energy or excreted. Both result in excretion of nitrogenous waste by the fish.

(b) Lipids

Lipids are a group of compounds with wide-ranging functions in the body (Jobling, 1993). Phospholipids are the main constituents of cell membranes; sphingomyelins are present in brain and nerve tissue; waxes and fats can be metabolized for energy; sterols are important components of the endocrine system, especially reproductive hormones. These are all chemical forms of fatty acids. In the context of this book it is primarily the fats that are of interest as these are the main source of energy in fish. They contain more energy per unit weight than any other biological compound (metabolizable energy: protein = 4.1 kcal g^{-1}; fatty acids = 9.2 kcal g^{-1}; carbohydrate = 3.5 kcal g^{-1} (Hepher, 1988).

Fatty acids are commonly classified according to the degree of unsaturation of the carbon chain of which each molecule is made (Hepher, 1988; Jobling, 1993). Each chain consists of a series of carbon atoms bonded together, the chain being of different length in different fatty acids. The differing physical and chemical properties of the fatty acids depend on the length of the carbon chains and the number and position of double bonds (Table 4.5). Saturated fats have no double bonds between the carbon atoms in the chain, and consequently have a high melting point, like margarine. Less saturated fats have one or more double bonds. The more double bonds, the lower is the melting point. Olive oil contains a high proportion of mono-unsaturates, i.e. fatty acids with one double bond, and vegetable oils have a higher proportion of polyunsaturated fatty acids (more than one double bond) and are therefore 'thinner' in consistency.

Fish, in contrast to mammals, have a high level of polyunsaturated fatty acids (PUFAs) in their tissues. PUFAs remain fluid at the low temperatures to which fish tissues may fall (fish being cold-blooded) and thus retain the activity of the cellular membrane transport mechanism. The PUFA group also contains the highly unsaturated fatty acids (HUFAs). Marine fish are a rich source of HUFAs and this is significant for the formulation of low pollution diets, as will be discussed in later sections.

Table 4.5 Examples of fatty acids illustrating methods of classification, and the effects of carbon chain length and degree of unsaturation on melting points (from Jobling, 1993)

Shorthand formula	*Structure*	*Melting point (°C)*
Saturated fatty acids		
16:0	$CH_3\ (CH_2)_{14}\ COOH$	63
20:0	$CH_3\ (CH_2)_{18}\ COOH$	76.5
24:0	$CH_3\ (CH_2)_{22}\ COOH$	86
Unsaturated fatty acids		
18:1 (n–9)	$CH_3\ (CH_2)_7\ CH = CH\ (CH_2)_7\ COOH$	13.5
18:2 (n–6)	$CH_3\ (CH_2)_4\ CH = CH\ CH_2\ CH = CH\ (CH_2)_7\ COOH$	–5
18:3 (n–3)	$CH_3\ CH_2\ CH = CH\ CH_2\ CH = CH\ CH_2\ CH = CH\ (CH_2)_7\ COOH$	–11
20:4 (n–6)	$CH_3\ (CH_2)_4\ CH = CH\ CH_2\ CH = CH\ CH_2\ CH = CH\ CH_2\ CH = CH\ (CH_2)_3\ COOH$	–49.5

(c) Carbohydrates

Carbohydrates are a group of chemically related compounds which include the sugars, starches, gums and cellulose. They can be utilized for energy in fish, either upon absorption or after long-term storage as glycogen in the liver. For most fish, particularly omnivores, carbohydrates form the greatest component of the diet. However, not all fish can fully utilize all the carbohydrate ingested. Omnivores, such as carp, can use the majority of carbohydrate ingested, but carnivores, such as salmonids or yellowtail, can only use about 25% of it (Jobling, 1993).

In formulated fish feeds, therefore, the proportion of carbohydrate is kept relatively low, as excess amounts will only be excreted and contribute to pollution. There are no essential carbohydrates and fish can survive on diets free of carbohydrate.

(d) Digestibility

A proximal analysis of a feed will indicate what proportions it contains of protein, lipid and carbohydrate. Not all of these components will be available for the fish to use for metabolism and growth as it may not be able to digest all ingredients of the diet. The reasons for this are the physical or chemical composition of the feed or environmental factors. The digestibility of feedstuffs is an important consideration in fish nutrition and pollution control: feedstuffs with low digestibility supply less nutrition to the fish and create more waste.

(e) Sources of dietary energy

For fish, the energy value of protein is higher than that of carbohydrate and it is more efficient for them to use protein as a source of energy for catabolism in preference to carbohydrate. The source of this protein may be either the diet, or metabolism of body tissues if the dietary intake of energy is insufficient. However, for a food production system this is an inefficient use of an expensive resource. It has been known for many years that by providing an abundant energy source, in the form of dietary lipid, the catabolism of dietary protein will be reduced, allowing more efficient use of the protein supplied. The protein level in the feed can therefore be reduced. (In fact, there is evidence that the extra lipid included in the diet leads to increased fat deposition if the level of protein in the diet is kept high.) This phenomenon is known as protein sparing, and forms the basis for the formulation of high energy diets (Johnsen and Wandsvik, 1991).

Returning to the principle elements of low pollution diets: high digestibility ensures less waste, as does an optimal nutrient balance. All the essential amino acids, for example, have to be present in the right proportions for the fish to gain the maximum nutritional value from the food. A food that cannot be utilized because an essential nutritional factor is in short supply is excreted by the fish. A high concentration of energy in the feed is achieved by including a high proportion of lipids in the feed. This ensures that as much as possible of the protein supplied is used for growth. If the protein were to be metabolized for energy instead, the amount of nitrogenous wastes produced by the fish would be greater.

This discussion of nutrition and feed formulation has been based on the knowledge and practices of the salmonid farming industry but the basic principles of nutrition apply to all fish and aquatic invertebrates. More specialist aspects, like protein sparing, are not as well researched in other species but have been demonstrated in a number of other fish species and in species of shrimp.

4.5.3 Husbandry factors

In addition to feed formulation, fish husbandry is an important component of any feeding strategy for waste minimization. This section discusses the husbandry factors that are important in this respect.

(a) Feeding methods

Feeding methods are an important husbandry factor in efforts to reduce the amount of waste generated in fish farming operations. Feeding method encompasses:

- method of presentation;
- timing of presentation;
- physical nature of the feed.

The primary objective in the process of feeding is to maximize the proportion of the food ingested by the fish and, from the environmental and economic point of view, to minimize the proportion wasted. In this respect feed quality has a number of criteria:

- pellet size
- proportion of dust
- stability in water
- sinking rate
- palatability.

The method by which the feed is distributed to the fish is also very important. The principal factors here are frequency of feeding, dispersion of the feed over the surface of the culture unit, and changes in appetite (due to weather, water quality, temperature, etc.).

(b) Feed delivery

The strategy adopted for delivering the feed to the fish can have a significant effect on the amount of feed wasted and the efficiency of feed utilization by the fish, and is therefore highly relevant to the control of pollution. Too little research has been carried out in this field to give hard and fast rules, but included here are the details of some recent experiments in order to illustrate the important aspects of feed delivery.

The experiments described here were conducted to investigate the effect of demand feeding on fish growth and feed use. In the first set of experiments, rainbow trout in floating cages were fed automatically by timer-controlled feeder, which delivered a portion of feed pellets at pre-set intervals, or by demand feeder (Alanärä, 1992a). The principle of a demand feeder is that certain fish behaviour stimulates a release of food. The unit in this case used a small rubber sensor suspended in the water, which triggered a release of food upon detecting a certain level of feeding activity (the fish would actually bite it, thus activating the sensor). It is worth noting that pendulum-type demand feeders, which release food in response to movement of the pendulum, can release food unnecessarily in rough weather when used on floating cages.

Two feeding strategies were used with the demand feeders: restricted and unrestricted demand feeding. In the latter case the fish had unrestricted access to food during daylight hours, simply by triggering the sensor. The restricted regime was based on the use of feeding tables to calculate the optimal daily feed quantity for the stock (with respect to the efficiency of conversion of food fed to fish growth): the fish were

Table 4.6 Specific growth rates (G_w) of rainbow trout (initial weight 1.1–1.2 kg) expressed as % per day and feed conversion (FCR) within different feeding regimes (modified from Alanärä, 1992a)

Feeding regime	G_w	*FCR*
Timer-restricted	0.72	1.36
Demand feeding – restricted	0.87	1.08
Demand feeding – unrestricted	0.93	1.49

able to feed on demand until this quantity was used up, after which no more feed pellets were available until the following day.

The specific growth rate and feed conversion ratio (FCR) are shown in Table 4.6. Quite clearly the demand-fed groups of trout out-performed those fed by timer-controlled feeder. Of the two demand feeder groups, the unrestricted feeding group grew faster, but at the expense of poor FCR. Those fed using the restricted demand feeding regime still grew well, but used very much less food.

It was found that in this experiment too little food was given during periods of hot weather, even when based on the optimal feed ration formula. It is essential, therefore, to have accurate information with which to determine the optimal daily feed ration if this technique is to work successfully. In order to achieve this, it was suggested, fish farmers may have to devise their own feeding tables by modifying standard tables to suit the conditions at each site. This would require long-term monitoring of environmental conditions and growth performance.

A further series of experiments, designed to overcome this problem, indicated that an alternative approach might be of more practical use (Alanärä, 1992b). Four groups of rainbow trout, again in floating cages, were fed according to different regimes (Figure 4.6). One group was given unrestricted access to the demand feeder, whilst the other groups had their access time restricted to 3, 5 or 7 daylight hours per day. It was hoped that by restricting feed delivery to a set period, feed waste resulting from excessive feeding activity would be reduced (Alanärä, 1992b).

The feeding response in fish is triggered by a range of stimuli, including hunger, presence of food, and observation of other fish feeding (Alanärä, 1992b). Some of the feeding activity observed in the groups with unrestricted access to the demand feeder could be the result of innate behavioural responses triggered by, for example, the presence of food or other fish feeding rather than by actual hunger, so that food is released unnecessarily, resulting in feed waste and poor FCR.

By analysing records of biting activity on the sensor, Alanärä (1992b) found that the appetite of the fish decreased significantly after the first

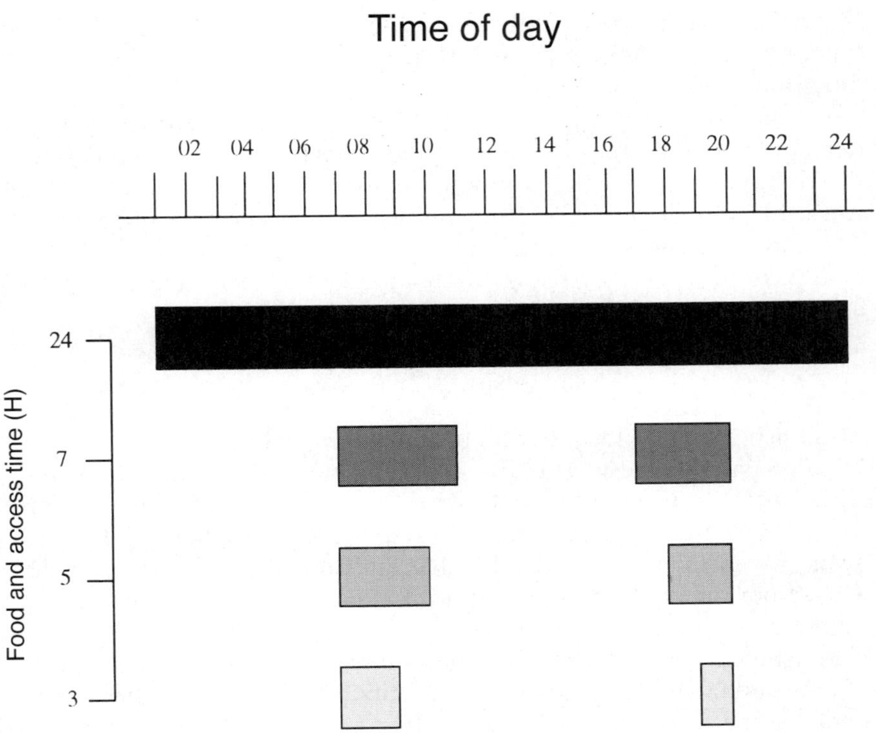

Figure 4.6 Fixed periods of available feeding hours in one unrestricted (solid bar) and three restricted feeding regimes. (From Alanärä, 1992a.)

2 hours of feeding activity during each feeding period. This suggests that they were becoming satiated and did not need further access to food until the contents of their stomach had been digested. This is borne out by the fact that by the end of the experiment fish in all three restricted regimes were showing good growth (1.1% per day) and good FCRs (1.0, 1.1 and 1.2 for the 3, 5 and 7 hour feeding regimes, respectively).

In a further experiment in this series, the effect of the energy content of the diet on food intake in a demand-fed stock of rainbow trout was studied (Alanärä, 1994). Many animals are able to adjust their intake of food so as to maintain a desired level of dietary energy. If fed on low energy diets, they eat more; if fed on high energy diets, they eat less. This experiment was designed to see if trout were able to adjust their activation of the food release sensor on the demand feeder so as to increase their intake of energy in response to the rise in temperature.

It was found that the trout could do this up to a point – at higher temperatures their bite activity increased. However, they apparently were not able to adjust their feeding level to the optimum for the energy content of the diet and the temperature. At 5°C trout grew at near their theoretical maximum at that temperature (1% per day). At 15°C, the theoretical maximum growth is 3.3% per day but the trout only achieved a daily growth rate of 1.9%.

The authors suggested that in order to optimize growth using a demand feeder of this type, it was necessary to vary with temperature the amount of food delivered upon actuation of the trigger. The optimal reward level will vary with temperature, feed energy content, and stock size and age. It was suggested that this could be set at about 0.1 g kg^{-1} of fish biomass at 5°C, rising to about 0.2 g kg^{-1} per trigger actuation at 15°C.

(c) Physical characteristics of the feed

Trash fish and offal were traditionally used in fish farming, and still are in some parts of the world. This source of nutrition suffers from some significant shortcomings. Firstly, it does not stay fresh for long; and secondly, some of the nutritionally important components – the fatty acids, for instance – quickly degrade and lose their nutritional value. Also, from the environmental point of view trash fish and offal can be highly polluting.

Moist diets were developed so that some of the nutritional deficiencies of offal and trash fish could be remedied by the inclusion of other components to the feed. Moist feed is a little more stable than minced trash feeds.

Dried feeds are much more stable than moist diets. Drying acts as a preservation technique, preserving the nutritional value of the feed ingredients. It also makes the feed much more stable, physically, in the water and consequently less polluting.

Extruded feeds are even more stable than compressed pellet dried feeds. Their 'aerated' structure allows the manufacture of floating and slow-sinking pellets which can be used to extend the presentation time of the pellet to the fish. This increases the probability that a fish will eat it, rather than the pellet falling to waste at the bottom of a pond or below a fish cage. The development of the extrusion process was an essential prerequisite to the development of today's low pollution diets. They can be manufactured with high inclusion levels of lipids. Their stability minimizes the production of feed dust during transport and storage, and the amount of wastage once the pellet has been broadcast over the water.

(d) Water quality

Attention to water quality is vital in order to derive the full benefit of low pollution diets, and this is another area where correct husbandry is very important in the reduction of pollution by dietary control. The main limiting factor is the availability of oxygen to the fish. Trials with rainbow trout in Denmark (Pedersen, 1987) have shown the following.

- Above 6 mg O_2 l^{-1}, feed consumption is independent of dissolved oxygen concentration; whereas below this level consumption reduced.
- Growth is reduced at maximum feed consumption when the oxygen concentration is less than 7 mg l^{-1}.
- Growth is reduced at 20% of the maximum feed consumption when oxygen concentration is less than 5 mg l^{-1}.
- Critical oxygen concentration probably increases with increasing energy content of the feed.

Stressed fish will consume more oxygen and if the water quality is not optimal then the fishes' demand for oxygen, to fuel the stress responses, will increase. This will then reduce the O_2 that is available for the full metabolism of high energy diets. If this falls below the critical level, the result will be increased waste production.

4.6 CASE STUDIES

Case studies 4.1 and 4.2 illustrate the practical benefits, with regard to pollution control, of using low pollution diets for intensive fish culture. The first illustrates the effect of feed formulation on nutrient discharge from a single fish farm, and the second the benefits on a regional scale of reduced nutrient discharge as a result of the widespread use of low pollution diets.

CASE 4.1

Alsted (1991) described a number of studies into the impact of low pollution diets on wastewater quality from trout farms. The studies were undertaken in Denmark and aimed to evaluate three aspects of feed formulation:

- Protein sparing – the effect of high lipid/low protein level on fish growth and carcass composition. The significance of protein sparing from a practical point of view is that protein is the primary source of nitrogen in fish feeds, and therefore of nitrogen in fish farm wastewaters. It is therefore desirable to reduce the input of protein to the system to a minimum, whilst maintaining an acceptable rate of growth in the fish. Altering the protein and lipid content of the feed may have an effect on flesh quality – another important consideration for the producer.
- The effect of different levels of non-protein nitrogen (NPN) on growth rate and pollutant load. In addition to the nitrogen bound up in

proteins, the feed will contain NPN (e.g. creatine, trimethylamine, trimethylamine oxide, taurine and ammonia). This is an important component of the diet, as 9–18% of the nitrogen in fish muscle is NPN (Alsted, 1991). Thus NPN may also have a significant impact on nitrogen excretion, depending on the quality and source of NPN in the diet.

- The effect of different inclusion levels of high digestibility carbohydrate on growth and liver size. Salmonids are not able to deal with high levels of carbohydrate in the diet, and excess is stored as glycogen in the liver (Hepher, 1988). Carbohydrates with excellent binding properties are required in high energy diets, which means that they are gelatinized and highly digestible. This characteristic may lead to fish health problems.

The chemical composition of the diets used in these experiments is given in Table 4.7, and fish growth in terms of weight gain per unit of nitrogen discharged is illustrated in Figure 4.7. The contrast between diets D and E (D being a lower protein/higher lipid diet resulting in higher growth, and E vice versa) illustrates the significant effect of protein sparing on nitrogen excretion. In this case, production of fish could be increased by 50% without increasing the nitrogen loading, or conversely the nitrogen output could be halved without loss of production. The higher lipid content of the diet was reflected in higher fat levels in the carcass, mainly around the gut and belly flap – the lipid content of fillets rose from 3.3% in diet E to 4.7% in diet D.

Diets produced from fish meals with different levels of NPN showed no significant difference with respect to discharges of NPN.

Table 4.7 Chemical composition of diets given to rainbow trout (from Alsted, 1991)

Component	*Diet designation*					
	A	*B*	*C*	*D*	*E*	*F*
Protein (%)	38.10	36.00	33.40	28.40	46.70	28.40
Lipids (%)	18.60	24.20	28.50	37.90	23.60	37.90
Carbohydrate (%)	27.30	25.40	24.00	20.50	11.20	20.50
Phosphorus (%)	0.82	0.78	0.75	0.64	1.00	0.64
Calculated metabolizable energy* ($kcal\ kg^{-1}$)	3845	4175	4379	4830	3950	4830
Calculated metabolizable energy contribution* (%)						
Protein	38	34	30	23	46	23
Lipids	39	47	52	64	48	64
Carbohydrate	23	19	18	13	4	13

*Using metabolizable energy as: 3.9 $kcal\ g^{-1}$ protein, 8.0 $kcal\ g^{-1}$ lipids, 3.2 $kcal\ g^{-1}$ carbohydrate

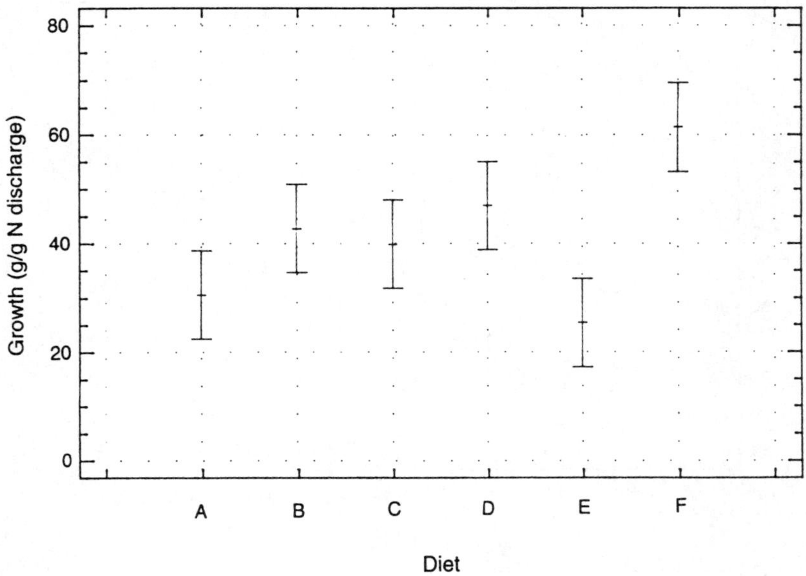

Figure 4.7 The effect of protein sparing on farm production is that the production of fish per unit of nitrogen discharged can be significantly increased (50% from diet E to diet D in this study), or the total discharge of nitrogen can be reduced significantly whilst maintaining the same production level. (From Alsted, 1991.)

Measurements of phosphorus discharge showed that this parameter varied significantly, depending on the source of fishmeal. The results of this trial indicated that by careful selection of the fishmeal it may be possible to 'choose' between excess P (that portion not assimilated by the fish) bound in the feed or dissolved in the water (Figure 4.8). This has implications for the formulation of diets for pond-reared and cage-reared fish. A considerable proportion of solid wastes settles out within the confines of the fish pond and can be removed mechanically, and so phosphorus bound up in solid form can be removed from the system before reaching the receiving waters. In cage rearing systems it may be preferable that a higher proportion of the P discharged is in the dissolved form. Alsted (1991) suggests that this would aid dispersion and prevent localized build-up.

The principle conclusions of this study are as follows.

- The discharge of nitrogen and phosphorus from fishfarms can be reduced considerably by altering the composition of the diet.
- Excess phosphorus in trout diets can be shifted between water-soluble and particulate fractions by selection of fishmeal.

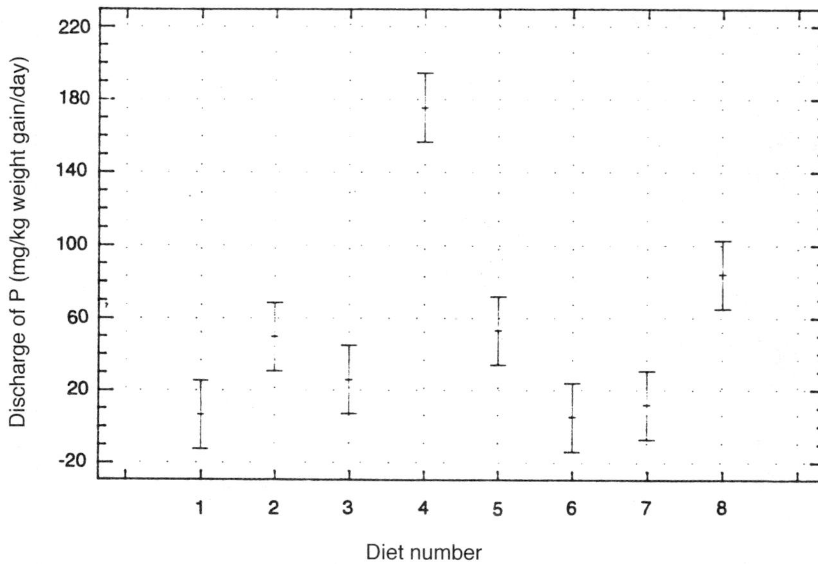

Figure 4.8 Discharge of total phosphorus by rainbow trout given diets containing different fishmeals. (From Alsted, 1991.)

CASE 4.2

Carlsson (1991) analysed national statistics to estimate the effect of the introduction of low pollution diets to the Swedish trout farming industry between 1980 and 1990.

The aquaculture industry in Sweden is dominated by the production of rainbow trout. The production figures for 1989 are summarized in Figure 4.9a. In Sweden pelleted fish feed contained about 1.2–1.4% phosphorus, until low pollution diets were introduced in 1986. These initially had a phosphorus content of 0.9–1.0% but by 1990 contained as little as 0.8% phosphorus. The effect of the use of low pollution diets on phosphorus emissions from trout farms (including both cage and land-based farms) is shown in Figure 4.9b. Phosphorus emission (kg t^{-1} net production) is relatively stable at 29 kg t^{-1} until 1985, after which it falls quite dramatically to 15 kg t^{-1}.

Figure 4.9c indicates that although the relative emission of phosphorus had fallen, the total loading of phosphorus on the environment had risen during the last three years of the survey. This was due to the increase in national production of fish during that period (Figure 4.9d).

The conversion to low phosphorus diets is certainly beneficial, as the risk of localized eutrophication will be reduced, but in sensitive ecosystems – such as lakes and semi-enclosed marine embayments – the fall

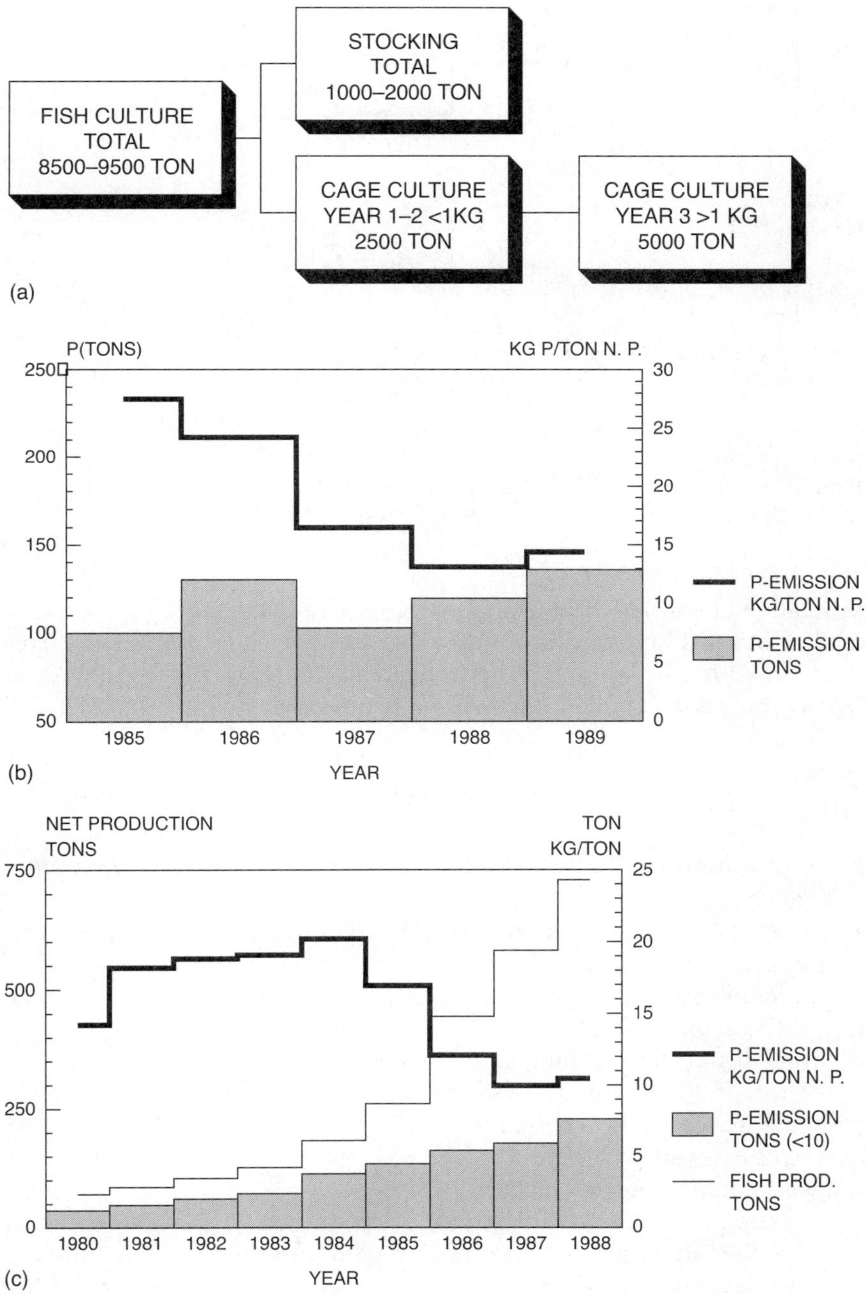

Figure 4.9 Analysis of Swedish fish farming data. **(a)** Estimated net production, 1989, with rainbow trout as dominating species. **(b)** Emissions of phosphorus 1985–1989 (ton fish = net production). **(c)** Net production of large rainbow trout and emissions of phosphorus in Kalmar county, 1980–1988.

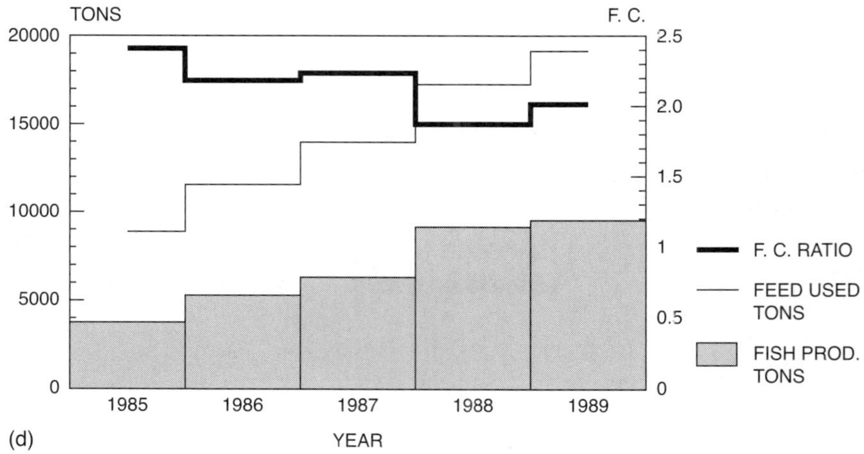

Figure 4.9 **(d)** Estimated Swedish fish production, used amounts of feed, and feed conversion ratio, 1985–1989. (From Carlsson, 1991.)

in relative emissions in phosphorus may be outweighed by an increase in the total loading of phosphorus due to increasing production. This illustrates the importance of a strategic approach to the regulation of aquaculture development and will be further discussed in Chapter 9.

SUMMARY

- Waste minimization is an effective strategy for environmental management.
- Strategies for waste minimization can be broadly classified into two groups:
  - planning and design;
  - operational management.
- Strategies at the planning stage include:
  - environmental impact assessment;
  - predictive environmental modelling;
  - culture system design.
- Operational strategies include:
  - strategic approaches to the use of chemicals;
  - feeding strategies (e.g. use of low pollution diets);
  - appropriate husbandry (e.g. feed delivery systems, water quality control).

CHAPTER 5

Wastewater reconditioning: physical methods

Wastewater from intensive aquaculture is inevitably contaminated with various by-products of the farming process – even with waste minimization strategies in place. If the water is to be re-used or if the impact on the receiving waters is to be reduced, these contaminants must be reduced to acceptable levels. The contaminants can be classified as follows.

- Solids
  - waste feed;
  - faecal material, mortalities, scales, mucus, etc.
- Dissolved substances
  - organic compounds;
  - metabolic wastes, i.e. ammonia, CO_2;
  - pharmaceutical and other chemicals.
- Pathogens.

In addition, oxygen will have been removed by the stock for respiration.

Wastewater reconditioning can be achieved by the use of either physical or biological treatment systems, or a combination of the two. Physical systems will be discussed in this chapter; biological systems will be dealt with in Chapter 6.

5.1 SOLIDS REMOVAL

Solids are conventionally removed by physical screening or by sedimentation, or a combination of the two processes (Cripps and Kelly, 1996).

5.1.1 Screening

Solid wastes from aquaculture systems are relatively small particles that are easily broken up, and microscreens are commonly used in their removal. Three main types of screen filter are used in aquaculture:

- rotating drum filter;
- rotating screen;
- fixed screen.

All these filters work on the same principle. They have a fine mesh screen, typically 60 microns for fish farm effluents, which strains solid particles, such as waste feed and fish faeces, out of the water. The small mesh size means that the screens clog quickly, and they need some mechanism for cleaning so that they continue to work properly (Lawson, 1995). Case studies 5.1 and 5.2 illustrate their means of operation, and their efficacy for solids removal.

CASE 5.1: UNIK® ROTARY FILTER

Bergheim *et al.* (1993) studied solids removal from a shore-based salmon on-growing unit using a UNIK rotary filter and a pilot sludge dewatering system, also by UNIK. The latter consisted of a microscreen minifilter and a sedimentation unit. The fish were supplied with a commercially prepared feed with the following composition:

- Total dry matter: 900 g kg^{-1}
- Total nitrogen (Kjeldahl-N): 66–70 g kg^{-1}
- Total phosphorus: 10–11 g kg^{-1}
- Metabolizable energy: 17–18 MJ kg^{-1}.

The average FCR during the trials was 1.2 kg feed kg^{-1} weight gain.

The design of the filter is illustrated in Figure 5.1. Two circular mesh screens are placed in the wastewater channel, one a coarse pre-filter (150 micron mesh size in this case) and one a fine filter (normally a 60 micron mesh, as in this case). As solids are entrained by the mesh the screens clog, and the water flow is reduced. To overcome this problem each screen rotates slowly, and is washed with a jet of water. The wash-water is collected in a trough and transferred to a settling tank where it is dewatered to leave behind a concentrated sludge.

Of the water quality variables measured, particle-dominated ones such as suspended dry matter (SDM) and total phosphorus showed the greatest variations in concentration and removal efficiency by the filter screens (Table 5.1). Of the dissolved substances, 25–50% of the soluble reactive phosphorus (SRP) was removed by the filter, whereas none of the total ammonia nitrogen was so removed. The authors believe that the SRP fraction was probably removed by adsorption on to the large number of particles on the filter screen.

Other studies have shown that microsieves in general have a relatively low removal efficiency of solids when these are at relatively low concentrations in the wastewater (2–5 mg SDM l^{-1} before treatment) (Bergheim

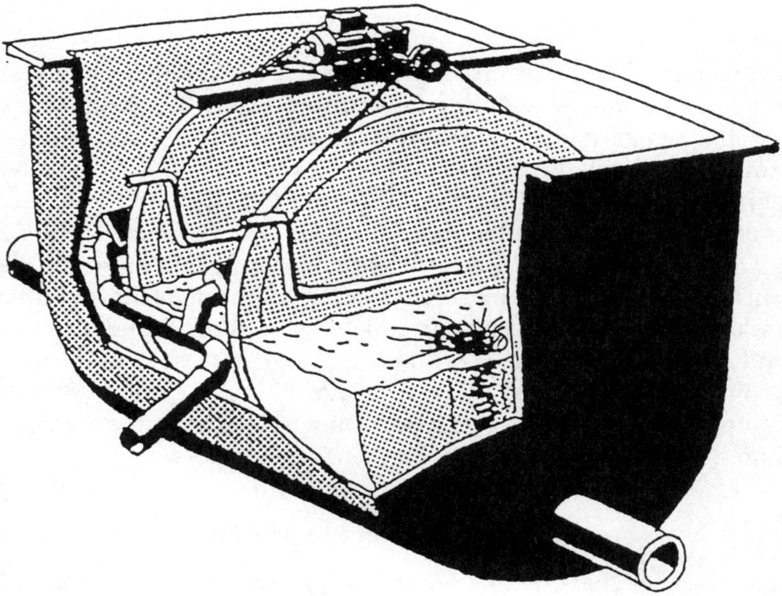

Figure 5.1 UNIK® rotary filter. (From Bergheim *et al.*, 1993.)

et al., 1993). At much higher concentrations of solids, commonly experienced during cleaning operations in fish tanks, the treatment efficiency was improved to 80–100% for SDM and total phosphorus. The authors claim in this study that the removal efficiency of the UNIK microscreen filter exhibits a higher removal efficiency than average, but the data is very variable and is presented without statistical analysis and should not be used to draw any firm conclusions.

Cripps and Kelly (1996) quote a study (in Norwegian) by Ulgenes which indicated a broadly similar range of treatment efficiencies to those described above for the UNIK microscreen filter.

Table 5.1 Wastewater treatment efficiency of UNIK® microsieves (150 microns + 60 microns) for Atlantic salmon shorebased seawater culture system (Bergheim *et al.*, 1993)

Water quality variable	*Treatment efficiency %*	
	Mean	*Range*
Suspended dry matter	68	19–91
Total phosphorus	63	47–84
Total nitrogen	17	7–32
Total ammonia nitrogen	0	< 0–5

CASE 5.2: TRIANGLE FILTER®

The Triangle filter has a fixed mesh screen set with a shallow slope through which the wastewater flows. Periodically a spray bar passes down the screen and its high pressure jets wash the screenings into a trough at the bottom, where they are collected for dewatering (Figure 5.2).

Makinen *et al.* (1988) tested the effectiveness of the Triangle filter 2400 for solids removal from fish farm wastewater. The filter screen had a mesh size of 65 microns. The effluent water was from a Finnish salmonid farm, using self-cleaning circular tanks. The authors found a reduction of phosphorus (TP as orthophosphate) across the filter of 78% on average. They qualify their results, citing methodological shortcomings, to the effect that phosphorus removal was anywhere between 40 and 80% – a significant improvement, nonetheless.

5.1.2 Pre-concentration of solids

Numerous trials have been undertaken to test the treatment efficiency of screen filters for aquaculture wastewater treatment. The results are very variable, even within one situation described. This is demonstrated in Table 5.1, where the efficiency of suspended dry matter removal ranges from 19 to 91%. This reflects two properties of aquaculture wastewaters:

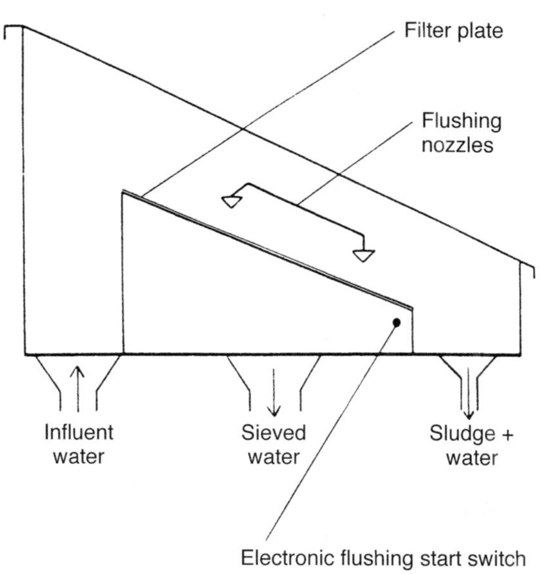

Figure 5.2 Basic elements of the Triangle filter®. (From Makinen *et al.*, 1988.)

- their high proportion of very fine particulates;
- their variability in composition, which is affected by management operations such as feeding and tank cleaning.

Screen filters are more efficient when the solids load is higher, and pre-concentration of solids by appropriate treatment of the wastewater stream is recommended. For example, solids concentrators are now available that can be fitted to the outlets of circular tanks; raceways can be designed with inbuilt sediment traps; or swirl-concentrators or hydroclones may be used. The sludge stream can then be treated using a microscreening device, then returned to the main flow. Thus the loading on downstream wastewater treatment systems, such as sedimentation ponds, or on the receiving waters is considerably reduced. At the same time wastewater treatment costs are reduced; the size of filter needed is proportional to the waterflow rate, not the solids loading.

5.2 SEDIMENTATION

Most intensive aquafarmers are familiar with the idea of a sedimentation pond as a method of treating aquaculture wastewaters. Normally, the residence time is the primary design parameter, but in this section it will be seen that this is an oversimplification of a complex process. Sedimentation units are also important components of recycle systems. They appear very simple at first sight, but their proper design and management requires some theoretical knowledge.

Flowing water has a certain level of energy. The faster the water is moving, the more energy it has – a fast-flowing stream will turn a waterwheel faster than a slow-flowing stream. Solid particles in the water are acted upon by two opposing forces: gravity and the energy in the water.

The more dense the particle, i.e. the heavier it is for a given size, the greater will be the gravitational force on it and the faster it will sink. Also, of particles of the same density, smaller ones will have a larger ratio of surface area to weight. Because the force exerted by the water on the particle depends on this ratio, smaller particles will stay in suspension for longer.

In practice this can be illustrated by considering a solid particle in a wastewater stream in which the speed of the water is gradually decreasing as the width of the channel increases. There will come a point when the force of the water acting on the particle is equal to and then less than the force of gravity acting on the particle. At that point the particle will begin to sink. It will still move downstream, but eventually it will sink to the bottom. More precisely, a particle of a given size and density will have a characteristic and constant settling velocity in water of a constant flow rate. This can be expressed mathematically (see Box 5.1).

In designing sedimentation units, the aim would be to make the length of the settling path as short as possible – that is, to have a very shallow settlement unit. Most intensive aquaculture systems continuously discharge large volumes of water and such a substantial flow put through a small, shallow lagoon or tank will have a very high velocity. To reduce the velocity to a level where solids will settle out, i.e. where gravity becomes the dominant force, it is necessary to increase either the depth of the pond or its surface area, or both, and so increase the residence time. The problem with increasing the depth is that it increases the settling path and, therefore, the settling time, which means that the particles may still be suspended by the time the water overflows from the sedimentation unit. However, if surface area is increased to increase residence time (by increasing either length or breadth, or both) the settlement time will be shorter at the same flow rate, and solids removal greater.

Therefore, from the design point of view it is the surface area that is the critical dimension (depth is also an important factor, but only within certain limits). Nothing will have been achieved if the result of making the tank shallow (in order to decrease the length of the settling path) is too high a flow velocity for settlement to occur.

Within these limits depth can be manipulated to increase the solids removal efficiency of settling tanks, especially with regard to smaller, slower-sinking particles. By experimentation it is possible to determine the settling velocity of the solids in aquaculture wastewater. This will not be one figure but a range of values, and the range will differ in each situation. Having determined the settling velocity of the slowest-sinking particles, the vertical distance they sink in a sedimentation tank with set dimensions can be calculated. Rather than increasing the dimensions of the tank to ensure that these particles settle out (assuming that this is necessary), the effective depth of the tank can be reduced without altering its flow characteristics. This could be done by installing a series of horizontal surfaces within the tank, placed so that the distance between them is less than the ideal tank depth for the smallest particles

BOX 5.1
SEDIMENTATION THEORY

The 'ideal' sedimentation basin, operating on a continuous flow basis and with a discrete suspension of particles, would have the following characteristics (Tebbut, 1992):

- quiescent conditions in the settling zone;
- uniform flow across the settling zone;

- uniform solids concentration as flow enters the settling zone;
- solids entering the sludge zone are not resuspended.

Using these assumptions its operational efficiency (% removal rate of particles with a given settling velocity) can be described thus. A discrete particle with a settling velocity v_0, and which just enters the sludge zone as it reaches the end of the tank, falls through a depth of h_0 in the tank retention time of t_0:

$$v_0 = h_0/t_0$$

But since $t_0 = V/Q$ (where V is volume and Q is flow per unit time), then:

$$v_0 = (h_0 \times Q)/V = (h_0 \times Q)/(A \times h_0)$$

where A = surface area of tank. Thus:

$$v_0 = Q/A.$$

Q/A is the 'surface overflow rate' and it follows that, for **discrete** particles, solids removal depends on this, not on the depth of the tank (Henderson and Bromage, 1988; Tebbut, 1992). The ideal settling pond to achieve sedimentation of particles of a given settling velocity (v_0) in a stream of given volumetric flow (Q) is defined by its surface area.

This relationship does not hold for **flocculating** particles. Here, the greater the depth of the tank, the more likely it is that agglomerations will increase in size as they settle. Larger agglomerations (flocs) settle faster than smaller ones, and so a larger proportion of the solids will be removed. In the absence of information to the contrary it is best to assume that solids in aquaculture wastewater do not flocculate.

A more normal situation is where a tank is fed with discrete particles of varying sizes. It is possible to determine the overall removal rate in the following way. The 'ideal' settling tank is designed to remove all particles with a settling velocity $\geq v_0$. Particles with a settling velocity $v_s < v_0$ will only be removed if they enter the tank at a distance from the bottom not greater than h, where $h = v_2 t_0$. Thus the proportion of particles with $v_s < v_0$ which will be removed is given by v_s/v_0 (Tebbut, 1992).

The settling velocity can be calculated from Stoke's law, given the diameter of particles, their specific gravity, and the kinematic viscosity of water (Lawson, 1995). Alternatively, empirically derived estimates of settling velocity can be used for sedimentation tank design (for methodology see Chen *et al.*, 1993b; Cripps, 1994).

in the wastewater. Thus the solids removal efficiency of the tank is maximized. It is difficult to clean accumulated solids from horizontal plates and so inclined plates or tubes are used instead. Settled solids agglomerate and tumble down the incline to collect in a sediment trap at the bottom. This should be designed so it can be flushed without emptying the tank.

In practice, solid removal is not a simple matter. A significant proportion of the solids in aquaculture wastewaters can be of particles of almost the same density as water and so their settling velocity is very slow. Very large facilities with very even (laminar) flow characteristics and no turbulence would be needed to ensure their removal. Conditions approximating these can be realistically achieved in small units for recycle systems, using inclined lamella plates or tubes. Even with these, it is unrealistic to expect effective removal of the very smallest solid particles. Henderson and Bromage (1988), in a study of 16 freshwater fish farms in the UK, concluded that it is unlikely that any sedimentation pond could achieve a reduction of suspended solids to below 6 mg l^{-1}. This is because of turbulent resuspension at the solid/liquid interface.

5.2.1 Problems with settlement ponds

In settlement ponds the real situation is a long way from theoretical ideals. Short-circuiting is often a problem. This occurs when, because of the shape of the pond, some areas are relatively stagnant whilst the water moves faster through other areas: the flow velocity may be higher than the design specification in some areas, and lower than necessary in others. The one does not compensate for the other, as most of the flow will pass through the short-circuit. There may be other factors that affect the pattern of flow through the pond, and consequently the settling efficiency. There may be density-induced currents, or temperature stratification, or wind-generated mixing. Obstructions in the pond, such as clumps of plants, will also affect the pattern of flow, and lead to suboptimal performance. The main effects of these factors are:

- to reduce the residence time of particles in the pond;
- to increase flow velocity to a level where settled particles may be resuspended.

Sedimentation ponds, therefore, differ very widely in their solids removal efficiencies, and rarely meet theoretical predictions. Rules of thumb which specify residence time only (not pond dimensions) are of little value. In the worst cases the solids loading can even increase within the sedimentation pond, because of resuspension of settled solids. Furthermore, the concentration of organic solids in the pond may exert a high oxygen demand, so reducing the dissolved oxygen concentration in the effluent.

The suspended bacterial population may also be high, and may itself significantly increase the BOD load on the receiving waters and affect natural microbial populations. Frequent desludging is essential.

The use of chemicals to enhance flocculation in sedimentation units is not practised in aquaculture wastewater treatment.

5.2.2 Recommendations for sedimentation pond design and use

In a study of existing sedimentation ponds, Henderson and Bromage (1988) found that most of the variation in their performance was accounted for by the mean fluid velocity in the pond. They suggest that settlement can be optimized by ensuring that the mean fluid velocity is less that 4 m min^{-1}, and preferably less than 1 m min^{-1}. Their findings differ from classic sedimentation pond theory in that the mean fluid velocity (length/residence time) is a function of pond depth as well as the volumetric flow rate (Q) and pond surface area (A). This might be a reflection of the greater complexity of processes to be found in real situations compared with those incorporated into theoretical models. It is also a reflection of the trade-offs between performance and other considerations (cost, space, etc.) which often have to be made in practice.

It would appear then that other parameters in sedimentation pond design, such as the inclusion of baffles, the use of length to width ratios, and residence time 'rules of thumb'(Henderson and Bromage, 1988), are of dubious value unless they bring about a reduction of the mean fluid velocity through the pond.

5.3 HYDROCLONES

The term hydroclone covers a class of solids separation devices that achieve separation by the generation of rotational flows: solids are separated faster than by gravimetric separation (sedimentation) alone (Lawson, 1995).

Hydroclones are commonly conical or cylindroconical units (Figure 5.3). Water enters tangentially, thus generating the rotational flow. This exerts a centrifugal force upon solid particles in the wastewater (Wheaton, 1977). Also a low pressure area – a vortex – is created at the centre of the unit which draws the clarified water through the central overflow; solids move outwards to the wall of the unit, then downwards. The greater the density of the particles, the less is the rotational speed necessary to generate the requisite centrifugal force to achieve solids separation. The conical section helps to direct the solids to the sludge removal outlet at the base of the unit. Five to 10% of the total flow through the hydroclone is drawn off with the solids and much of this can be returned to the system by dewatering the sludge.

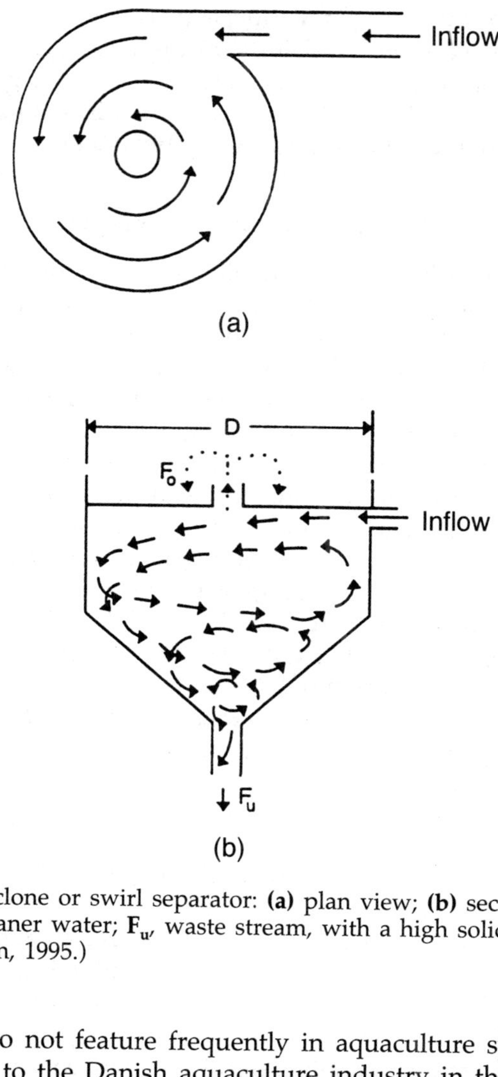

Figure 5.3 Hydroclone or swirl separator: **(a)** plan view; **(b)** sectional view. $\mathbf{F_o}$, overflow of cleaner water; $\mathbf{F_u}$, waste stream, with a high solids concentration. (From Lawson, 1995.)

Hydroclones do not feature frequently in aquaculture systems. They were introduced to the Danish aquaculture industry in the early 1980s (Figure 5.4) as a space-efficient alternative to sedimentation ponds. Trials demonstrated that a swirl-concentrator (as this particular design was described) of 9 m diameter (64 m^2 in area) could achieve solids removal comparable to that of a 500 m^2 sedimentation pond (Warrer-Hansen, 1982).

As with sedimentation ponds, swirl-concentrators cannot achieve high removal rates of very fine solids. These have a density very similar to that of water, and it is not possible to generate sufficient centrifugal force to separate them. Hydroclones have also been installed as pre-filters, upstream of biofiltration units, in recycle systems (Scott and Allard, 1983). They require far less space and can match the performance of a conventional sedimentation tank.

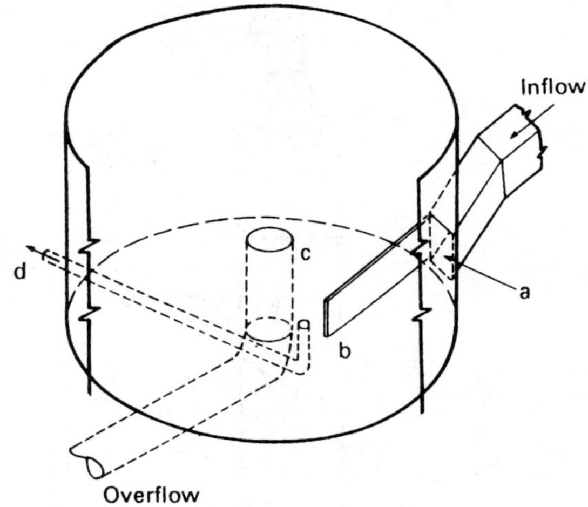

Figure 5.4 Swirl concentrator: **a**, wastewater inlet; **b**, flow deflector; **c**, overflow of clarified effluent; **d**, outlet for separated solids. (From Warrer-Hansen, 1982.)

5.4 FLOTATION

Flotation, also known as foam fractionation (Chen *et al.*, 1993a), is a process whereby low density particles and dissolved organic molecules are entrained in a stream of air bubbles and concentrated in a surface foam, which is removed from the flotation chamber and dewatered separately (Figure 5.5) (Lewis, 1980). This process will be familiar to fish-keeping hobbyists as protein skimming, but the process is used in aquaria primarily for the removal of dissolved organic substances.

In addition to this function, flotation can also be very effective in the removal of small and slow-sinking particles, which are difficult to remove by sedimentation and screening. Despite the increasing use of flotation in municipal wastewater treatment, its application to the treatment of fish farm wastewater is not widespread.

Flotation units are ideally suited to inclusion in an aquaculture water treatment system. Aeration is frequently used, especially in recycle systems, for fish and biofilter respiration and also for degassing (of ammonia and CO_2), especially in recycle units. In addition, flotation removes the surfactants – seen as films or scum on the water surface – which can reduce oxygen transfer rates. These are particularly prevalent where oily, high energy diets are used. A bubble column can also be useful for water transfer as an airlift pump.

Flotation is the transfer of a solid from the body of a liquid to the surface by means of bubble attachment (Lewis, 1980). It is a gravity separation process in which the attachment of the gas bubbles to the solid

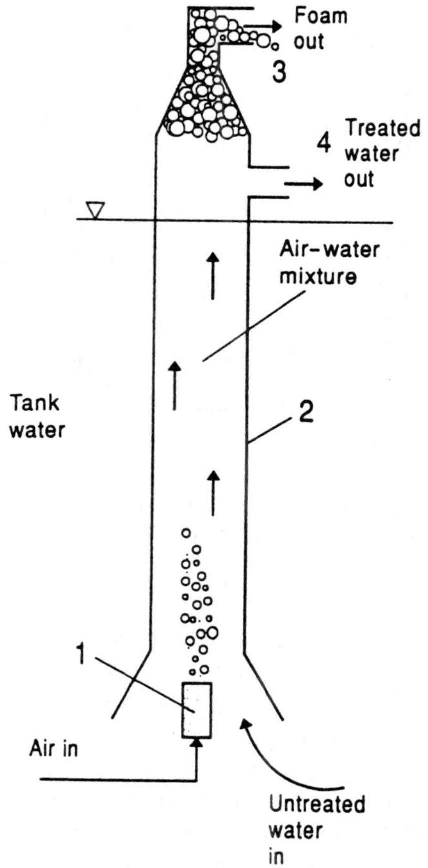

Figure 5.5 Co-current flotation unit (▽ water level). (From Lawson, 1995.)

particles lower their apparent density below that of the water, so that the bubble/solid agglomerate floats to the surface. The attachment of the particle to the bubble depends on the surface-active properties of the particle. Thus proteins with a hydrophobic portion are attracted to the air–water interface of the air bubbles in the flotation column, and are concentrated in the resulting foam. Non-surfactant proteins are not attracted in this way, and are therefore not removed from the culture water by this process.

The main variables in the performance of flotation units are (Weeks *et al.*, 1992; Chen *et al.*, 1993a,b; Lewis, 1980):

- gas bubble size (the smaller the bubble size, the more efficient is the unit for particle removal);
- diffuser depth (determines the contact time between the bubble and the water column, and thus influences particle removal rate);

- gas flow rate (affects the removal rate and the foam overflow height);
- water flow rate (affects the contact time);
- foam overflow height (determines the concentration of particles in the foam condensate and is itself a function of gas flow rate and design of foam collection device).

Air is normally used as the gas. There are several different types of flotation process and they are characterized by the method used to produce the air bubble. Electroflotation is based on the production of fine bubbles of hydrogen and oxygen by passing a direct current between two submerged electrodes. Cost considerations make the application of this technique to aquaculture unlikely.

5.4.1 Dispersed-air flotation

There are two types of dispersed-air flotation system in use (Lewis, 1980). In foam flotation, air is introduced to the flotation chamber by being forced through a diffuser (usually a porous ceramic or plastic material). This is a simple method, but the air bubbles produced are large (> 1 mm diameter) and the system is therefore less effective for solids removal.

Froth flotation entails the use of a fast-rotating impeller to produce the bubbles from the incoming stream of air. The high shear forces involved in this method lead to the break-up of delicate solid particles and so it is not well suited to the treatment of aquaculture wastes.

5.4.2 Dissolved-air flotation

In dissolved-air flotation, the air bubbles are not introduced into the water but are formed in the water by the reduction in pressure of water saturated with air. The bubbles so released are very small – typically 20–100 μm in size (Lewis, 1980). The systems of particular interest to fish farm wastes are microflotation and pressure flotation.

(a) Microflotation

In this system the wastewater is aerated as it is passed down a U-shaped shaft about 10 m deep. As the water travels down the shaft the hydrostatic pressure increases, allowing an increase in the amount of air dissolved in the water. As the water rises up the other shaft of the U, the hydrostatic pressure is reduced and the saturation pressure point falls, and air is evolved from the water as very fine bubbles. The amount of air available for flotation is restricted by the depth of the shaft.

(b) Pressure flotation

Air is dissolved in water under pressure, then evolved as fine bubbles as the pressure is reduced. There are three methods for dissolving air into the water:

- full-flow pressure flotation, in which the whole influent stream is pressurized and aerated;
- split-flow pressure flotation, in which a fraction of the flow is pressurized and aerated;
- recycle-flow pressure flotation, in which clarified effluent is aerated under pressure and introduced into the flotation chamber.

Recycle-flow pressure flotation is probably the method best suited to fish farm wastes. Part of the clarified wastewater is recycled under pressure to supply the air bubbles for the clarification process. The advantage over the full-flow and split-flow methods is that none of the solid particles are subjected to the shear forces resulting from the pressure reduction required to release the air bubbles. The disadvantages of this method over the full-flow method are that the flotation tank has to be larger as a result of the increased flow due to the recycling, and that in order to supply the same amount of air the recycle water must be subjected to higher pressure (Lewis, 1980).

5.4.3 Summary

Microscreen filters for aquaculture typically operate with 60 μm screens, and are unable to remove very fine suspended solids (less than 30 μm), which may make a significant contribution to the BOD loading of wastewaters. The potential of flotation systems for removing solids that would pass through a typical microscreen is of considerable interest. Dispersed-air flotation and dissolved-air flotation are inherently simple processes which could easily be adapted to aquaculture wastewater treatment. However, there are few data on the performance of flotation units for aquaculture wastewater treatment. Such as they are, the data are encouraging, but a great deal of development work still needs to be done.

5.5 SLUDGE DEWATERING AND DISPOSAL

Sludge from filtration devices usually has a high water content. There are a number of commercial dewatering devices available to concentrate the solids further and they use a combination of screen filtration and sedimentation. Other devices are available, such as reverse osmosis and ultrafiltration, but these are unlikely to be cost effective for the foreseeable future.

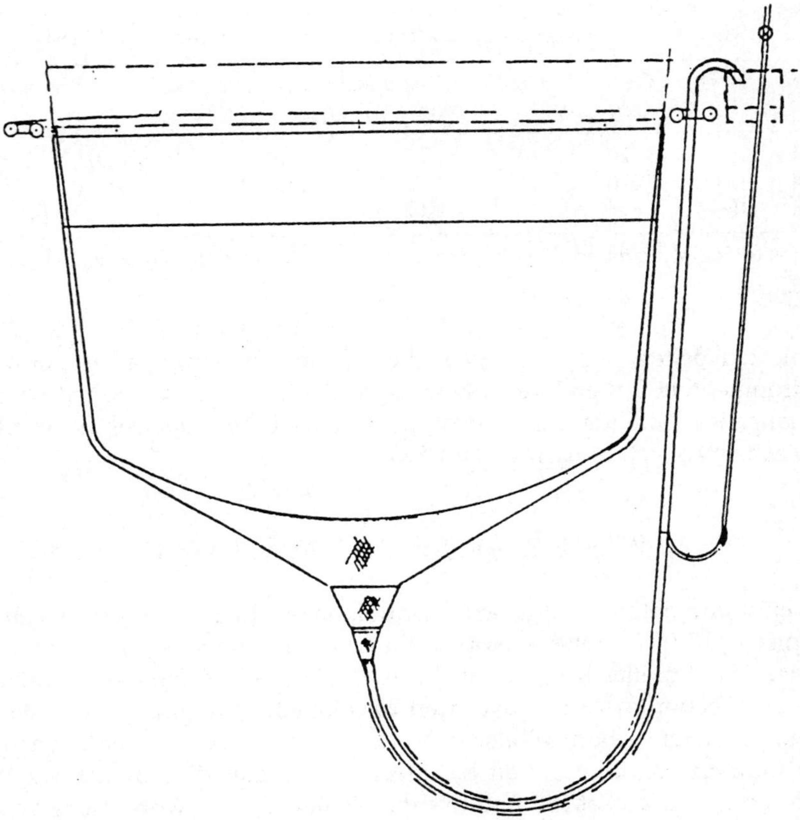

Figure 5.6 'Lift-up' feed collector, which collects surplus feed from under net pens and transports the feed back to the surface. The return pipe is shown weighted. (From Røed, 1991.)

The most practical method for aquaculture sludge dewatering is drying or partial drying in open tanks or ponds, then disposal to land as a fertilizer. Care should be taken to prevent seepage of nutrient-rich water (this will have a very high BOD) into the groundwater.

5.6 INTERCEPTION OF SOLID WASTES FROM FLOATING CAGE AQUACULTURE

Although not widely used, there are collection devices designed to intercept waste feed pellets from floating cages (Figure 5.6). An outer net placed around the fish cage directs sinking pellets to a central collection funnel (Røed, 1991) and a small compressor at the surface sucks the waste pellets to a collection device. This unit was designed to collect

Table 5.2 Comparison of oxytetracycline concentrations in incoming and effluent wastewater from a salmon hatchery in Ireland (Smith *et al.*, 1994)

Date	*Time*	*Oxytetracycline used (g/day)*	*Filter inflow conc.**	*Retentate conc.**	*Filtrate conc.**
12/4	16.00	1270	< 0.02	12.8	< 0.02
12/4	17.00	1270	< 0.02	11.6	< 0.02
20/4	16.15	616	< 0.02	10.0	< 0.02

*Concentration of oxytetracycline given in mg l^{-1}. Limit of detection = 0.02 mg l^{-1}.

medicated pellets and so prevent them from remaining in the natural environment; it is, therefore, only used intermittently. Reduction of water exchange is a potential disincentive to the use of these devices as permanent waste collectors (Cripps and Kelly, 1996).

5.7 WASTE PHARMACEUTICAL REMOVAL

The environmental concerns about antibiotic use in aquaculture systems are discussed in Chapter 3. Norway has strict regulations governing the release of antibiotics from aquaculture operations into the environment. Waste collection systems have been developed for cages, as described in the previous section. Whilst it is not normal practice in the aquaculture industry, antibiotics can be effectively removed from the wastewater by filtration. Table 5.2 shows the results of some work carried out in Ireland (Smith *et al.*, 1994). Clearly the filter (of 50 μm nominal porosity) retained substantial quantities of oxytetracycline.

5.8 ACTIVATED CARBON

Activated carbon works by adsorption; that is, it attracts and binds molecules to its surface. The carbon granules have an incredibly high surface area and therefore a high absorptive capacity. This is not limitless and they do need periodic recharging (by heating to high temperatures) or replacement (Lawson, 1995).

Activated carbon is usually used to remove organic compounds from water, and so improve colour and remove distasteful smells – an important factor in high quality domestic water supplies. They are also effective in removing some metal ions from water. Acid-treated activated carbon will also remove ammonia.

Activated carbon is not widely used in the treatment of aquaculture wastewaters and one reason is cost, which is high when large volumes have to be treated. Another is that activated carbon in wastewater streams with high BOD or ammonia-N or nitrite-N will soon be colonated by a population of bacteria, which reduce the absorptive capacity of the carbon granules. However, it may prove useful for specialized applications – for example, studies in the USA have achieved 99% removal of malachite green from fish hatchery wastewaters using activated carbon filtration systems (Marking *et al.*, 1990).

5.9 ION EXCHANGE

In this process the wastewater is introduced to a substrate that has many electrically charged ions adsorbed to its surface (Lawson, 1995). As the wastewater passes over this substrate, certain ions in the wastewater are exchanged for the ions on the substrate, because they have a stronger affinity for the substrate. Substrates are either manufactured resins or artificially occurring zeolites. There is a wide range of zeolites, each type having a characteristic selectivity (Pond and Mumpton, 1984). Clinoptilolite has a high affinity for ammonium and has been used successfully for ammonia removal from freshwater aquaculture systems. As with activated carbon, its absorptive capacity is limited and the granules need periodic recharging, which is achieved by flushing with sodium chloride solution – clinoptilolite will not work in seawater.

Cost and practical difficulties with large-scale installations preclude the widespread use of zeolites for aquaculture wastewater treatment but there may be specialist applications – ammonia removal in seasonally operated recycle units in fish hatcheries, for example (start-up time makes the use of biofilters problematic).

5.10 PATHOGEN ELIMINATION FROM WASTEWATERS

There are cases when disinfection of aquaculture wastewaters is necessary for pathogen removal. The most practical methods for aquaculture purposes are oxidation using ozone, and ultraviolet sterilization.

5.10.1 Ultraviolet sterilization

When properly administered, ultraviolet irradiation is effective against all aquaculture pathogens, with the possible exception of some viruses (Liltved *et al.*, 1995). It works by interfering with the DNA in cells (Lawson, 1995). The efficacy of UV sterilization depends on:

- UV bulb wattage, age and cleanliness;
- distance between the bulb and the target organism;
- species and individual characteristics of the target organism;

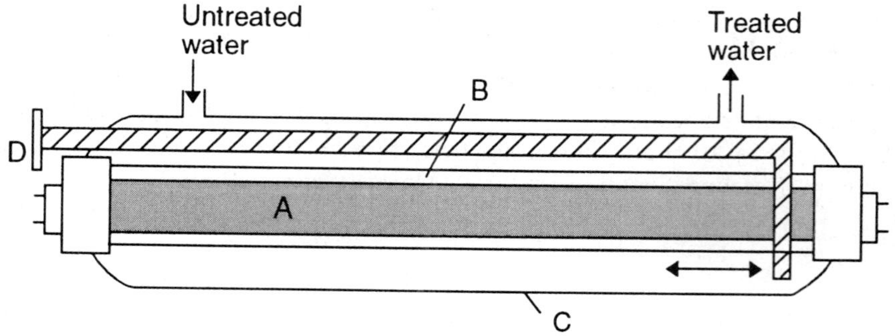

Figure 5.7 Typical quartz-jacketed UV sterilization unit: **A**, UV bulb; **B**, quartz tube; **C**, stainless steel photo reactor; **D**, pushrod for cleaning. (Redrawn from Lawson, 1995.)

- duration and intensity of exposure;
- clarity of the culture water.

Modular UV sterilization units specially designed for aquaculture are commercially available (Figure 5.7). The most important factors governing their sterilization efficiency are the exposure time and light intensity. The design and operational factors of the culture system affecting these are discussed below.

- **Light intensity**. The intensity of light falling on the target organism is a function of bulb wattage and water clarity. Water absorbs UV light very quickly and so it does not penetrate very far. It is recommended that target organisms pass within 25 mm of the bulb to ensure effective sterilization. Many commercial units operate on much shorter distances than this. The turbidity and colour of the water, if high, will significantly reduce the penetration of UV light and therefore reduce the dose received by the target organisms; it is important to have some form of pre-filter, certainly for solids. Colour is not usually such a problem in aquaculture systems.
- **Exposure time**. This is determined by the flow rate of water through the UV sterilization unit: it should be kept relatively low for effective sterilization. The optimal flow rate will depend on the design of individual units.

Light intensity and exposure time together constitute the dose of UV irradiation.

One important point to remember is that UV sterilization is only effective on organisms that are exposed to the UV light. Even with an effective UV sterilizer in operation, significant populations of micro-organisms can exist elsewhere in the system, especially in pipes (Lawson, 1995). For this reason ozone sterilization is often recommended.

5.10.2 Ozonation

Ozone (O_3) is a highly unstable form of oxygen. As such it readily dissociates and is a very powerful oxidizing agent. This oxidizing power, which burns cell membranes and inactivates enzymes, makes ozone a very good disinfectant. It also has other benefits for use in aquaculture systems, including removal of BOD and reduction of suspended solid loads by oxidation of organic molecules. Ozonation units are often advocated for pre-filtration in biofiltration systems: reducing the influent BOD loading ensures more efficient nitrification (Lawson, 1995). In this application it is important to de-gas the influent stream before it reaches the biofilter, where any residual ozone would seriously affect biofiltration. The factors that affect the efficacy of ozone sterilization are:

- concentration of dissolved organics in the water;
- concentration of particulate organics;
- type and composition of inorganic ions;
- pH;
- temperature.

In contrast to UV sterilization units, which are affected by similar factors, pre-filtration is not normally undertaken to remove dissolved and particulate organics. Instead the dose of ozone is increased to compensate if their concentration is high. This is because ozonation is effective for their removal from the system, as well as for sterilizing the water.

Ozone is so unstable that it has to be generated on site, and the ozone generator is a major component of any ozonation system. The ozone is generated as a gas by means of an electric arc, and then injected into the water. Minimum contact times are recommended for different applications, and a separate reactor vessel in which this takes place is advisable. Thus most of the ozone can be confined to just one part of the culture system. This is necessary because ozone is such a powerful oxidizing agent that it is toxic to the farm stock and to nitrifying bacteria, and it is important to keep the residual concentrations in the culture system as low as possible. Thus a means must be provided for the ozone to degrade or be removed from the water before the treated water enters the biofilter or culture tanks. The best method is to trickle the water through a packed tower (Figure 5.8) to strip the residual ozone from the water. Ozonation can be combined with flotation (Figure 5.9).

Ozone in the atmosphere is also toxic to humans, and attention should be paid to the health and safety aspects of ozone use. Manufacturers of ozonation systems should be consulted on this matter.

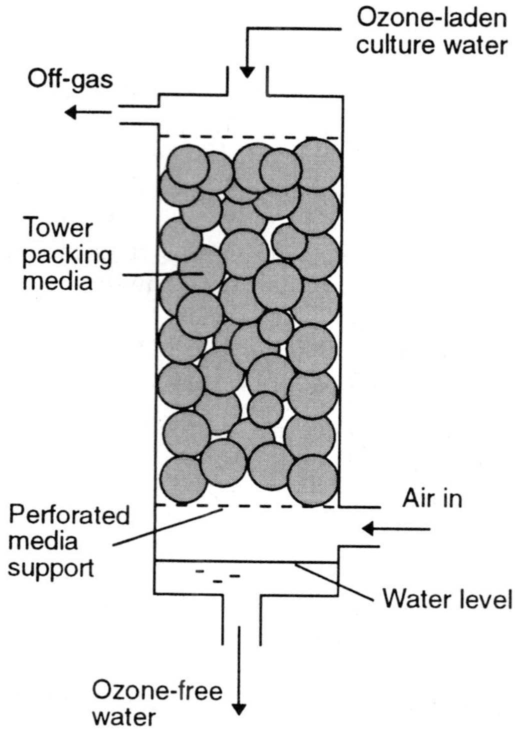

Figure 5.8 Packed column of the type used for gas stripping in aquaculture (ozone in this case). (Redrawn from Lawson, 1995.)

5.10.3 Filtration for pathogen removal

Cartridge filters are available with a small enough pore size to achieve the removal of all pathogens, including bacteria and viruses. These systems are expensive; they are also restricted to relatively low flow rates, and need comprehensive pre-filtration to prevent clogging.

Pressurized sand filtration may be used to achieve pathogen removal, especially of larger organisms. Its efficacy for bacteria and viral pathogens is unlikely to match that of ozone and UV sterilization.

5.11 OXYGENATION

The management of dissolved oxygen (DO) in the culture systems is important for the following reasons.

- Low DO decreases the efficiency with which the farm stock utilize artificial feeds, thus creating more pollution (Chapter 4).

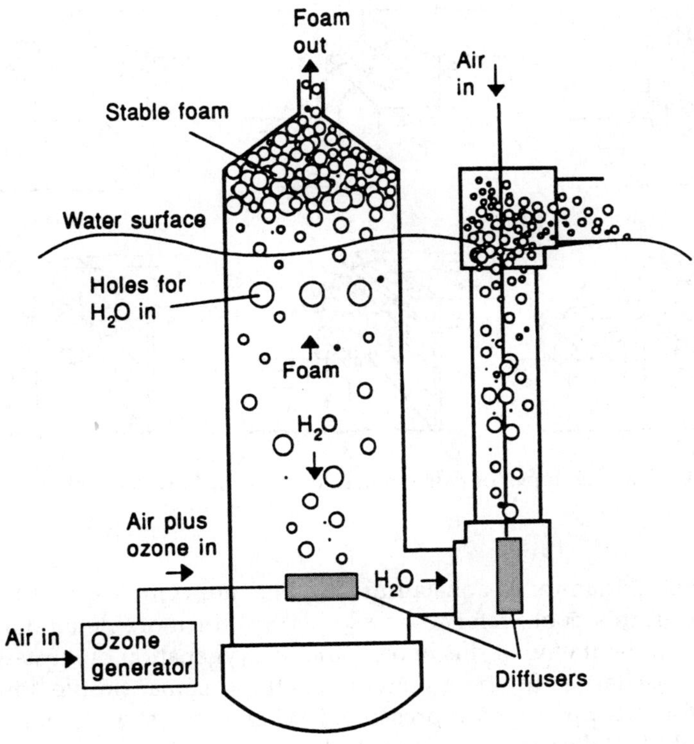

Figure 5.9 Combined ozonation and flotation unit. (From Lawson, 1995.)

- Low DO causes stress responses in many cultured species. This increases susceptibility to disease and may increase the discharge of chemicals (Chapter 3).
- Biofilters have a high oxygen demand, and will not work effectively where this demand is not met. This can increase levels of pollution (Chapter 6).
- Aquaculture wastewaters are usually deficient in dissolved oxygen. This must be remedied prior to discharge to prevent pollution effects on receiving waters (Chapter 2).

There are many off-the-shelf systems available to increase the DO concentration in aquaculture ponds, recycle systems and wastewaters (Colt and Tchobanoglous, 1981; Speece, 1981; Lawson, 1995). From the practical point of view they must be assessed with regard to:

- the efficiency with which the transfer of oxygen is achieved;
- their operating and capital costs.

The choice of system employed will depend on the specific design, operating and economic characteristics of individual culture systems.

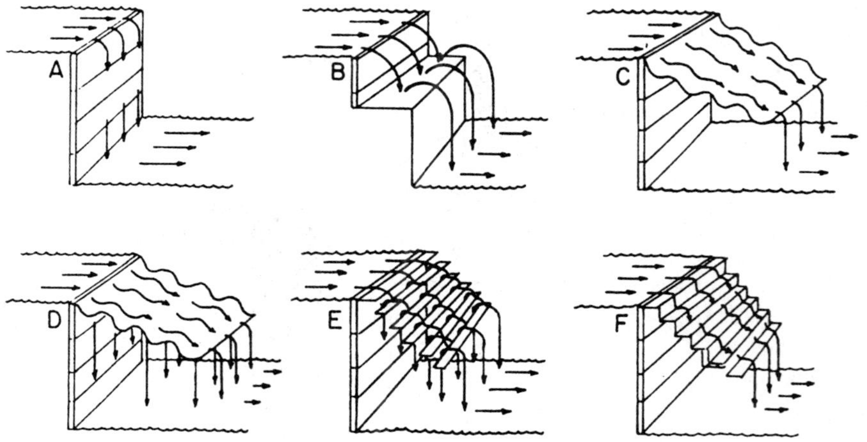

Figure 5.10 Various types of cascade aerators. (From Lawson, 1995.)

The oxygen dynamics of aquaculture systems are complex, and beyond the scope of this book; refer to Lawson (1995) for more detail. Of most relevance to the theme of this book is the re-oxygenation of wastewaters prior to discharge. In the following sections some simple low-cost devices for this purpose are presented. Whatever system is employed, care should be taken to ensure that turbidity is not increased as a result of oxygenation. This may occur through the resuspension of settled solids, or by erosion of channel banks due to turbulent waterflow generated by the device.

5.11.1 Cascades

Cascades are artificial waterfalls, and may have a wide variety of configuration (Lawson, 1995) (Figure 5.10). The basic principle is twofold:

- to break up the stream of water as much as possible so as to maximize the surface area for oxygen exchange;
- to cause the entrainment of air bubbles, and so aid oxygen exchange.

To this end, perforated plates, artificial riffles and weirs are incorporated into their design. The oxygenation efficiency of cascades is governed by:

- the efficacy of the unit in maximizing the surface area of the water stream and in achieving the entrainment of air;
- the vertical height of the cascade – the higher the cascade, the longer is the contact time for oxygen exchange and the more energy there is in the system for the entrainment of air.

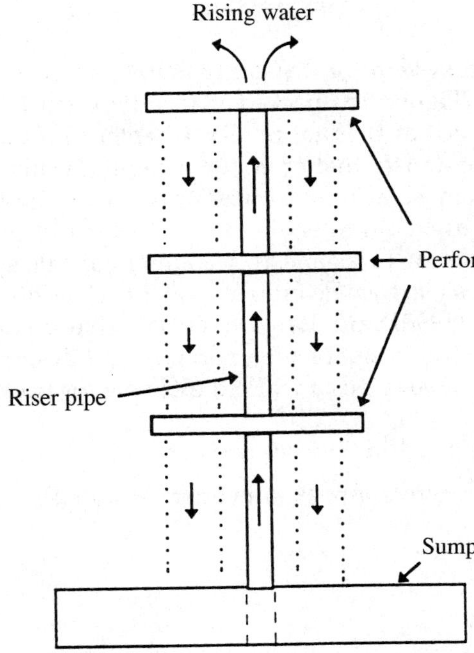

Figure 5.11 Type of perforated plate aerator. (After Lawson, 1995.)

5.11.2 Upwelling gravity aerator

This is a form of cascade often used on pond inlets, but equally applicable to piped discharges (Lawson, 1995). The wastewater upwells through a pipe then falls to the main water body through a series of perforated plates arranged around the pipe (Figure 5.11) and is aerated in the process.

5.11.3 Vertical pump systems

These comprise a submersible electric motor mounted vertically and driving an impeller that forces water into the air. A large surface area for oxygen exchange between the water and the air is provided, both by the formation of many water droplets and by the entrainment of air bubbles as the droplets fall back into the main body of water. These units are often mounted on floats. Another variation on this theme is a submersible pump that discharges through a spray bar. All are available in a range of sizes – the power rating of the motor giving an indication of its oxygenation capacity.

5.12 DEGASSING

The most widely used system for degassing is the packed column. This is usually a cylinder (Figure 5.8) packed with a high surface area/low void space material, such as biofilter media (Chapter 6) (Lawson, 1995). The media ensures the transformation of the incoming water into a very high surface area stream, so as to maximize the opportunity for gaseous exchange across the gas–liquid interface. Air or oxygen is introduced at the bottom of the column to establish a counter-current system, and gaseous exchange takes place as the influent water trickles down through the column. Watten (1990) and Lawson (1995) give comprehensive accounts of the theoretical basis for the operation and design of packed columns, and Watten (1990) gives a worked example for packed column design.

Packed columns are used in aquaculture for:

- nitrogen removal (where supplies are super-saturated);
- oxygenation;
- ammonia removal;
- ozone removal;
- carbon dioxide removal.

The most common applications are to the treatment of wastewaters in recycle systems (e.g. ammonia removal, oxygenation) or hatchery influent (e.g. nitrogen removal, ozone removal) or effluent streams (e.g. oxygenation).

SUMMARY

- A range of contaminants find their way into the wider aquatic environment from aquaculture systems. Physical treatment systems can be used for their removal.
- Physical treatment systems are suitable for the removal of:
  - waste feed;
  - faecal material;
  - pathogens.
- Solids removal systems include:
  - screening;
  - sedimentation;
  - hydroclones;
  - flotation;
  - interception (from cages).
- Pathogen treatment systems include:
  - filtration;
  - UV irradiation;
  - ozonation.

- Aquaculture wastewaters can also suffer from oxygen depletion and the presence of dissolved gases. Systems for the mass transfer of gases are briefly reviewed.

CHAPTER 6

Wastewater reconditioning: biological wastewater treatment systems

Biological treatment systems are primarily used for the removal of dissolved substances from the water, though some are concerned with the mineralization of organic solids. The systems are diverse in nature but can be classified into three categories:

- biofilters
- constructed wetlands
- bacterial augmentation systems.

6.1 THE BIOFILTER

The biofilter is the most common biological treatment system. It relies on the metabolic processes of microbial communities to remove dissolved organic compounds from the water and convert toxic dissolved inorganic nitrogen to less toxic forms.

To gain a better understanding of how biological treatment systems work it might be useful to ask how the natural processes of aquatic ecosystems are harnessed to achieve wastewater treatment.

6.1.1 Freshwater ecology

Aquaculture wastes, with the exception of some chemical residues, are not toxic (Chapter 2). They are damaging to the environment because of the changes they cause in the nutrient status of aquatic ecosystems. Indeed, aquaculture wastewaters are no more than relatively concentrated solutions of nutrients containing suspensions of organic and mineral solids.

In natural communities it is mainly the microbial community, as distinct from the macro-invertebrate, fish, or aquatic plant communities, that are best able to utilize these nutrients – hence the extensive bacterial and fungal growths in waters that are grossly overloaded by nutrient-rich wastewaters. Aquatic macrophytes and algae do have a role to play but they are not as efficient as the bacteria in using the nutrients directly and are usually out-competed in severely polluted freshwaters. In less polluted situations the increased supply of nutrients may promote periphytic algal growth and phytoplankton productivity (Carr and Goulder, 1990a,b).

The microbial community in a polluted river exhibits serial changes in species composition downstream of the wastewater discharge point. Improvements in water quality can also be detected and the two are very closely related (Gray, 1989) – one of the earliest river pollution classification schemes was based on the composition of the microbial community because this so closely reflected the water quality. Figures 6.1 and 6.2 (Gray, 1989) illustrate the basic elements of the saprobien system, developed in the early 1900s, and the microbial relations that underlie it. This system identifies four zones, defined by the dominant ecological processes within them:

- degradation
- active oxidation
- recovery
- 'normal functioning' (in relative terms).

As with most classifications, this one draws artificial boundaries in what is, in fact, a continuum. Examination of the critical indicators is much

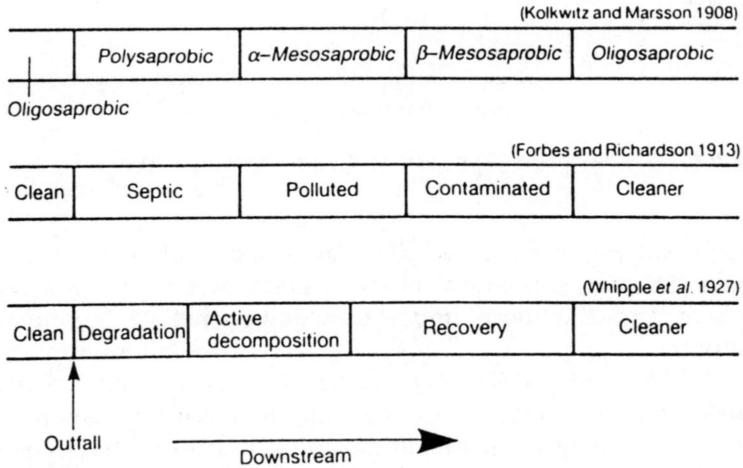

Figure 6.1 Comparison of the saprobien river pollution classification system (top) with other early descriptive systems. Those of Forbes and Richardson (1913) and of Whipple *et al.* (1927) illustrate the broad functional relationships of the saprobien system. (From Gray, 1989.)

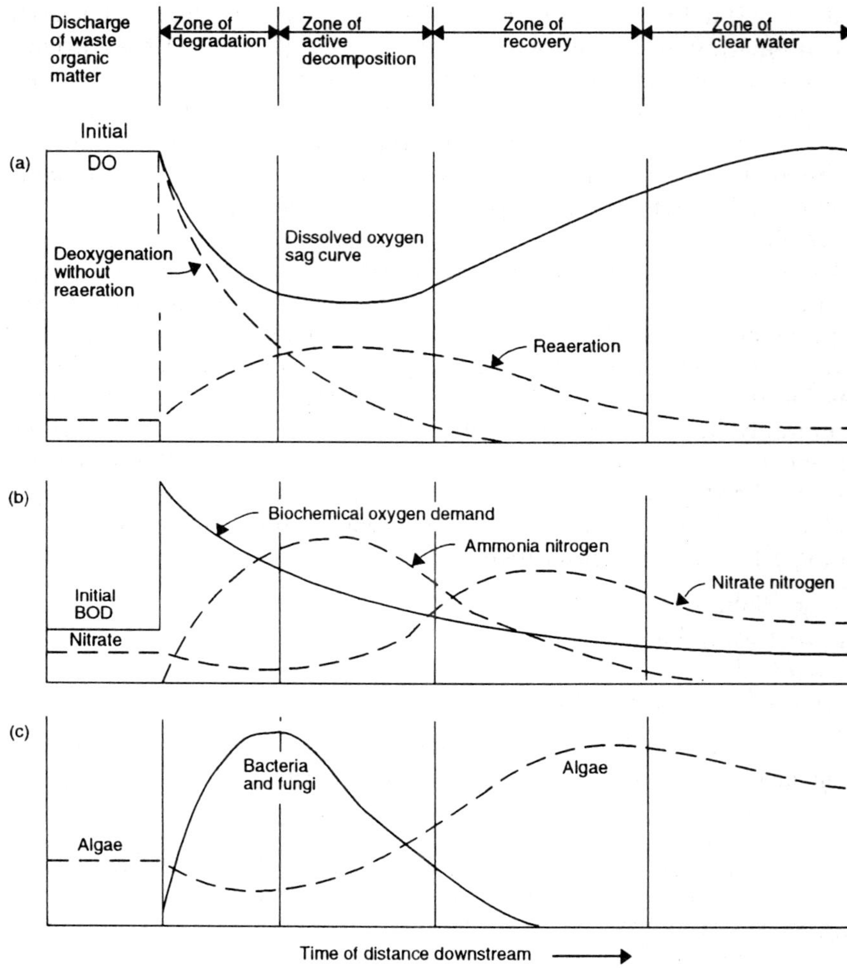

Figure 6.2 General effects of organic pollution in streams. (From Gray, 1989.)

more relevant. Figure 6.2 shows the relationships within the continuum, between the physicochemical characteristics of the polluted aquatic ecosystem and the more important components of the microbial community.

The discharge of nutrient-rich wastewater into a natural waterway stimulates a rapid increase in the aquatic bacterial population, which feed on the nutrients contained in the effluent. Most of the bacteria are species that need to attach to a suitable substrate (Gray, 1989). On our own scale of perception we tend to think of the surfaces of rocks, plants and logs as substrates. Bacteria certainly attach to these, but also of considerable importance are suspended solids as substrates for the attachment of bacteria. Most suspended solids are just visible to the naked eye, whereas bacteria are microscopic. A fish faecal strand is enormous relative to the size of a bacteria, and considerable numbers attach to it.

In situations where the nutrient loading is very high, large microbial growths may form attached to fixed surfaces. These growths are often referred to as sewage fungus, but are in fact a more complex association of heterotrophic slime-producing organisms, including both bacteria and fungi. The greatly increased microbial activity in these initial zones causes a significant decrease in the concentration of dissolved oxygen. The greater the level of pollution, the greater will be the microbial activity and the greater the reduction in dissolved oxygen. In severely polluted conditions the water is often devoid of oxygen.

Gradually, along the length of the river and as time progresses, the level of microbial activity declines as the level of dissolved nutrients is reduced, and the dissolved oxygen concentration in the water rises. This is recognized in the oxygen sag curve (Figure 6.2a), the depth of the sag representing the severity of the pollution and its effect on the river community. As the dissolved organic compounds are metabolized, so reducing their concentration in the water, the concentration of nitrogenous waste products in the form of ammonia-N increases. It is of note that concentrations of nitrate-N lag behind those of ammonia-N.

As the chemical characteristics of the water change from being a solution rich in organic carbon compounds, towards one having a relatively high concentration of inorganic nitrogen compounds, so the community changes. Heterotrophic bacteria, which use the dissolved organic compounds in the wastewater as a carbon source, are progressively replaced first by ammonia-oxidizing bacteria and then by photosynthetic organisms, both periphytic and planktonic algae (see Box 6.1).

The micro-organisms are not the only organisms present but they form the basis of the community. Preying or grazing upon these micro-organisms is a whole range of protozoa, flagellates, ciliates and rotifers, and then preying upon these consumers is a whole community of higher animals (Gray, 1989). What is present in each zone depends on the prevailing conditions. As evidenced in Chapter 2, the diversity of species in a community is considerably reduced in polluted conditions.

What has been described above is a typical ecological community of a polluted river – a 'polluted river ecosystem'. In the active zones in such an ecosystem, i.e. those in which the nutrients are being processed by the microbial community, the bacteria and fungi essential to this process are attached to all the available surfaces: to gravel, rocks, sticks and tree branches, even to the submerged parts of any surviving aquatic macrophytes.

To create an artificial polluted river ecosystem one would have to provide surfaces to which microbes could attach, and a constant supply

BOX 6.1
AUTOTROPHS AND HETEROTROPHS

In common with all living things, bacteria need essential nutrients of two types:

- those required to provide energy for growth and maintenance;
- those required to supply the chemical elements needed for biosynthesis (the manufacture of complex biological material).

Plants and some bacteria can meet their energy needs by utilizing sunlight, whilst most bacteria and all animals need to ingest energy-rich organic or inorganic compounds. Plants and those bacteria using inorganic compounds for energy and biosynthesis are known as **autotrophs**. Those animals and fungi using organic compounds to supply their energy and as a source of organic molecules for biosynthesis are **heterotrophs**.

Bacteria of both types are important components of ecosystems with regard to the processing of metabolic wastes of higher animals such as fish and crustacea. Products of animal metabolism include faecal wastes, which consist of solid or dissolved organic compounds. Dead plant material has similar characteristics and enters the microbial food web in a similar way to animal faecal waste. These organic molecules are used by heterotrophs as their nutrient and energy source.

Metabolic by-products also include inorganic nitrogen wastes resulting from the breakdown of proteins for energy. In the aquatic environment this is usually in the form of ammonia. Certain autotrophic bacteria are able to use this ammonia, which is toxic to many organisms, as a source of energy. In doing so they are able to convert it to less toxic forms of nitrogen and these are used by plants as nutrients.

Nitrification and denitrification

Some autotrophic bacteria use inorganic nitrogen as their primary energy source. *Nitrosomonas* sp. uses ammonia for this purpose and in so doing converts it to nitrite; *Nitrobacter* sp., in turn, converts nitrite to nitrate (Lawson, 1995). This process is known as nitrification. Denitrification, the conversion of nitrate to nitrogen gas, is effected by other bacteria only in anaerobic conditions.

of nutrient-rich water. If this was then arranged in a very compact form it could process considerable quantities of nutrients in a very small space (compared with a natural ecosystem). This is the basic concept of biological treatment systems for aquaculture wastewaters.

6.1.2 Ecology of the biofilm

In biofilters, bacteria attach to the surfaces provided by the filter media. They produce a mucous film which supports quite a diverse microbial community (Figure 6.3), including bacteria, fungi, protozoa and larger invertebrates (Gray, 1989; Bruce and Hawkes, 1983). This film has its own characteristic ecology, i.e. the relationships between the organisms present and the physical and chemical environment. The dynamics of this ecosystem is what ultimately determines the performance of the biofilter in reducing nutrient concentrations in wastewater.

The biofilm is composed of a jelly-like matrix secreted by the purifying bacteria, in which the bacteria themselves are embedded. This film is referred to as a matrix because it is more like a sponge than a solid mass; hence water can flow both over its surface and through it. Purification occurs via a number of processes (Bruce and Hawkes, 1983):

- Solids removal: fine and colloidal solids are flocculated as water passes through the matrix by extracellular polymers secreted by the micro-organisms. Flocculated solids, and those physically trapped, are digested by extracellular enzymes and absorbed as soluble matter by the micro-organisms.
- Protozoa and macro-invertebrates living on the film may ingest trapped or flocculated solids directly.
- Assimilation of soluble compounds: the two classes of dissolved matter (organic compounds and ammonia-N) diffuse into the film to be absorbed by the micro-organisms and oxidized or assimilated intracellularly.

The oxygen required for these processes reaches the cells by diffusion through the film. The waste products of the microbial activity, mainly CO_2 and mineral matter, diffuse through the film into the stream of water flowing over its surface. Gaseous products, notably CO_2, ultimately diffuse into the atmosphere.

Biofilm thickness

The thickness of the biofilm has important consequences for its efficiency in purifying the wastewater (Bruce and Hawkes, 1983). The level of microbial activity depends on the maintenance of an adequate supply of oxygen and either organic compounds or ammonia-N within the biofilm. This depends on diffusion from the wastewater into the biofilm

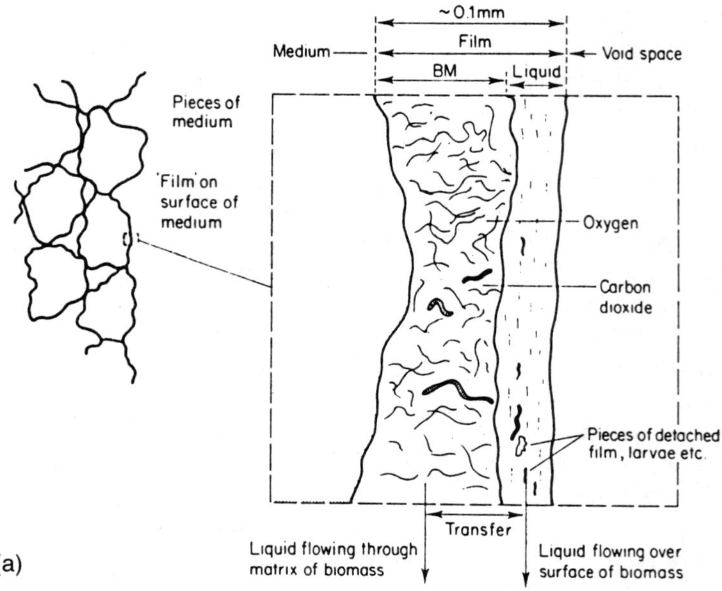
~0.1mm
Film
Medium
Void space
BM
Liquid
Pieces of medium
'Film' on surface of medium
Oxygen
Carbon dioxide
Pieces of detached film, larvae etc.
Transfer
Liquid flowing through matrix of biomass
Liquid flowing over surface of biomass

(a)

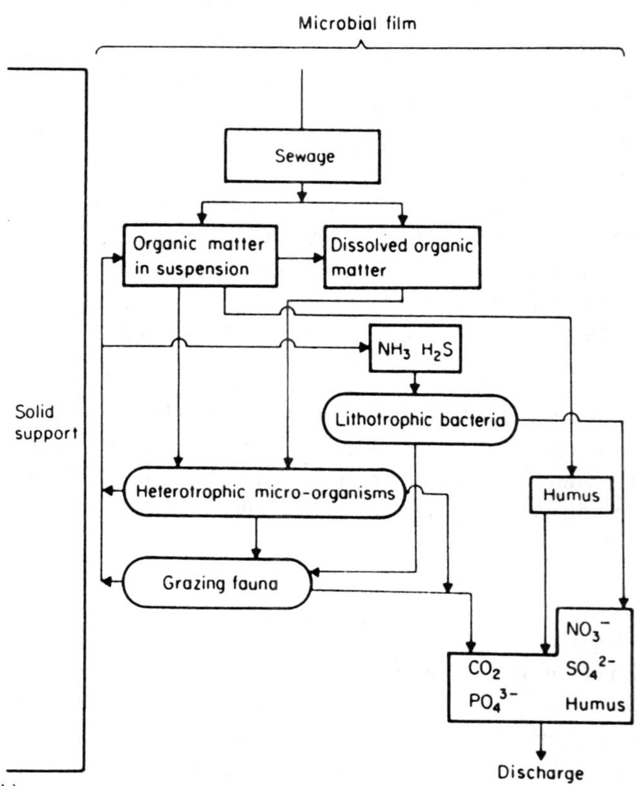
Microbial film
Sewage
Organic matter in suspension
Dissolved organic matter
NH3 H2S
Lithotrophic bacteria
Solid support
Heterotrophic micro-organisms
Humus
Grazing fauna
NO3−
SO4 2−
CO2
PO4 3−
Humus
Discharge

(b)

matrix. As these substances migrate through the depth of the film their concentration is progressively diminished by microbial activity. In a very thick film the nutrient or, more usually, oxygen becomes exhausted and anaerobic conditions prevail. Within the aerobic depth of the film, zonation may occur between heterotrophic and autotrophic bacteria (Bruce and Hawkes, 1983). Nitrifying autotrophs do not compete well with heterotrophic bacteria and fungi when organic matter is abundant, because of their slow rate of growth. Thus the outer region of the film in these conditions is dominated by heterotrophs. With increasing depth the concentration of organic matter decreases, as the microbes use it for their metabolism, and the concentration of ammonia-N consequently rises. Gradually, as the balance between organic matter and ammonia-N is reversed the microbial community becomes dominated by nitrifying bacteria (Figure 6.4). This zonation is not necessarily present, however. Depending on the nutrient characteristics of the wastewater the film may be heterotroph-dominated or autotroph-dominated.

The same change can be observed, on a much larger scale, as wastewater flows downward through a trickling filter. Near the surface the biofilm might be predominantly heterotrophic, depending on the characteristics of the wastewater. Towards the bottom of the filter, assuming it is an appropriate size for the nutrient loading that it receives, the biofilm might be dominated by nitrifying bacteria (e.g. Nijhof, 1995).

6.1.3 Implications for biofilter design and management

This knowledge of biofilter ecology can be applied to achieve optimal design and performance of biofilters. The implications are discussed in the following paragraphs.

Management objectives

It is important to decide what is required of the biofilter – either to remove carbohydrates (so reducing BOD), or to convert ammonia to nitrate. In aquaculture applications the reduction of BOD can be achieved mainly by solids removal systems, the dissolved component being not so high as to limit biofilter performance with regard to ammonia conversion (by competitive exclusion of nitrifying bacteria). This is an important consideration, as it may affect the choice of biofilter medium and the loading rate on the filter.

Environmental control

As with any ecosystem, the community of the biofilm is adapted to relatively stable conditions. Slow changes are tolerated but sudden ones are

---

Figure 6.3 Structure of **(a)** a microbial film and **(b)** the food web within it. (From Bruce and Hawkes, 1983, and Gray, 1989, respectively.)

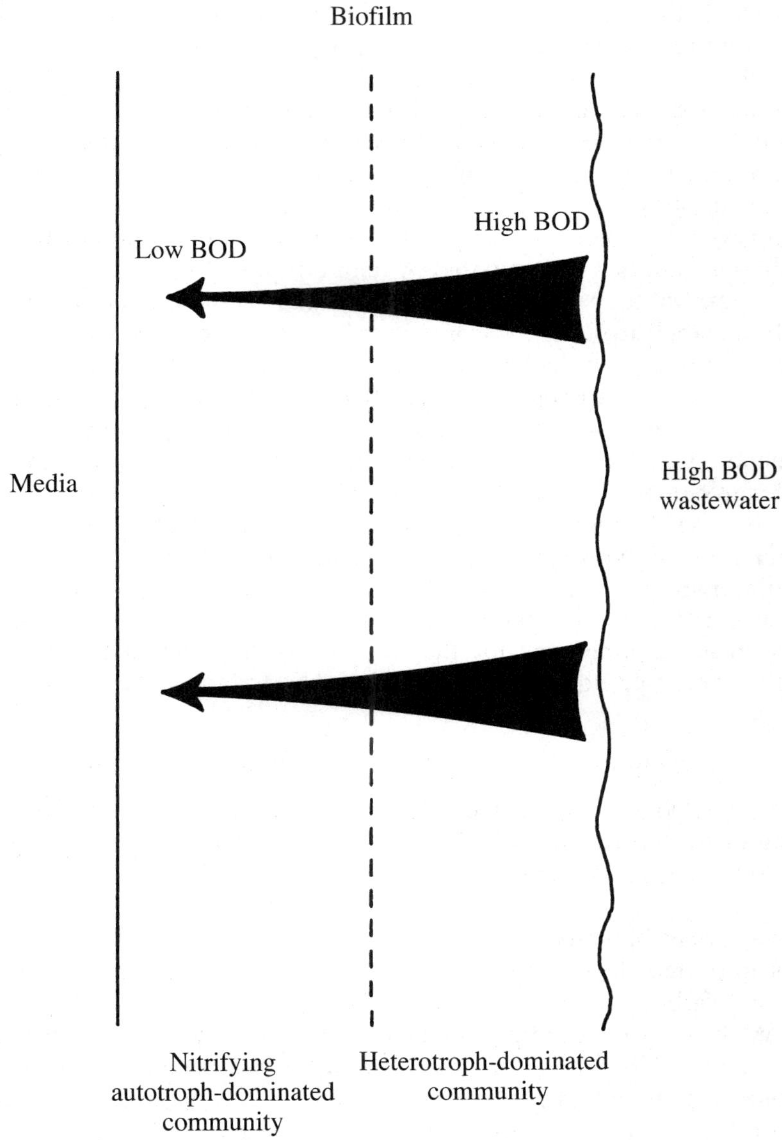

Figure 6.4 Zonation within the microbial community of a biofilm. The actual wastewater flow through a biofilm matrix would be much more diffuse than shown in this schematic representation.

not. Thus temperature and pH, and salinity for marine systems, should be kept as constant as possible.

Biofilters work best in total darkness. It is thought that light can inhibit the metabolism of *Nitrobacter* and *Nitrosomonas*.

Any changes in hydraulic loading rate or ammonia or nutrient loading rate should be introduced slowly so as to allow the system time to adapt. It is difficult to give specific recommendations, but a minimum of a few days should be expected. It would be advisable to have a back-up system in place to cope with any reduction in treatment performance during this time.

Pharmaceutical use

If biofilters are to be incorporated into an aquaculture system, then provision must be made for the treatment of medicated water by alternative means. Many anti-parasitic treatments and all antibacterials and antibiotics will harm the performance of the biofilter if medicated water is not diverted.

Loading rate

In the design of biofiltration systems for aquaculture the most important variable is the ammonia concentration in the wastewater from the culture unit. Knowledge of the ammonia production rate is essential and in this respect the mass balance approach recommended by Lawson (1995) is very useful.

Note that the ammonia production rate is ultimately determined by the fish feed input, not by the biomass of fish in the system. The two are very closely linked, but the feed input is the critical factor. The biofilter is there solely to process the wastes from the utilization of this food by the stock in the culture system.

Contact time

A minimum depth of 1.8 m is often recommended for trickling filters, to allow adequate contact time between the wastewater stream and the biofilm for effective treatment, but this advice is derived from experience with traditional sewage treatment biofilters using graded gravel as a substrate medium (Bruce and Hawkes, 1983; Tebbut, 1992). The optimal depth of bed for a trickle filter depends on a variety of factors, the most important being specific surface area of the substrate, BOD and ammonia loading rates, and hydraulic loading rate.

Denitrification

Nitrate can be removed from the water by the microbial process of denitrification, in which nitrates are converted to nitrogen gas. This process requires anaerobic conditions and is therefore not generally suitable for aquacultural applications.

Specific surface area

The specific surface area is a measure of the surface area of the filter substrate medium per unit of volume, expressed as $m^2\ m^{-3}$. The greater the specific surface area of the medium, the greater is the filtration efficiency with regard to filter volume. This must be offset against the potential reduction in efficiency due to clogging of the void space: generally, the greater the specific surface area, the smaller is the void space. In practice, most manufactured filter media are designed as open ring or cylindrical forms, which maximize void space and minimize the likelihood of clogging. Typically, these plastic media have specific surface areas ranging from 100 to 400 $m^2 m^{-3}$.

Aeration

Aeration can be used in submerged filters to ensure adequate oxygen supply to the biofilm. The use of aeration, or oxygenation, essentially converts a two-phase system (solid/water) into a three-phase one (air/solid/water). There is the additional cost of the aeration unit itself, so three-phase trickle filters are generally more desirable. Aeration in trickle filters can help to drive off ammonia from the wastewater (Chapter 5).

Pre-filtration

Pre-filtration is important, as high solid loading can cause clogging of the filter medium, both by direct accumulation and by the stimulation of excessive growth of biofilm or sewage fungus due to the high BOD.

Cleaning

Biofilter units may need cleaning from time to time, but this should not be frequent in well designed systems. During cleaning it is important not to remove all the biofilm from the filter medium if the system is to be restocked immediately. Light rinsing of the medium is sufficient. Units should be designed to allow decanting of settled solids without closing the system down.

Sloughing of the biofilm

One important feature of the biofilm with regard to managing biofilters is the constant turnover of the community. Parts of the biofilm are periodically washed off the surface of the medium. This may occur for either of two reasons:

- the water velocity is too high; or
- the biofilm has become too thick.

The latter situation will occur when the loading on the biofilter (especially in the case of carbohydrates) is high, encouraging excessive growth of the microbial community and of the biofilm.

This is not necessarily a problem. In fact, the constant turnover probably promotes removal efficiency. Populations progress through a series of stages as they become established in any environment (Ricklefs, 1979).

In the earlier stages of expansion the population increases in size rapidly. Later, as the food resources or space become limited, the individuals in the community have to compete with each other for resources, and so the growth rate of the population slows. Continuous sloughing of biofilm from the filter media ensures the continued availability of resources for growth, thus maintaining a vigorous population.

There are implications for management of biofilters which must be considered:

- Excessive biofilm growth can block the filter, especially where a filter medium with low void space is used.
- Provision must be made to remove the sloughed biofilm from the water leaving the filter. Commonly a sedimentation chamber is installed downstream of the biofilter for this purpose.

Biofilter start-up time

Like any biological community, that of the biofilter requires some time to become established – for the initial 'seed' community to multiply to its optimal size and community balance in relation to the nutrient composition of the wastewater. This is commonly referred to as the start-up time, during which the biofilter does not operate at maximum efficiency. If animal stock are introduced before the biofilter is conditioned, there may be problems with water quality and consequently poor fish health: it is preferable to stock the system after the microbial community has become fully established. The development of the microbial community in a new biological filter is best illustrated by reference to their effect on the water quality downstream of the filter (Figure 6.5). Initially the biofilter will have

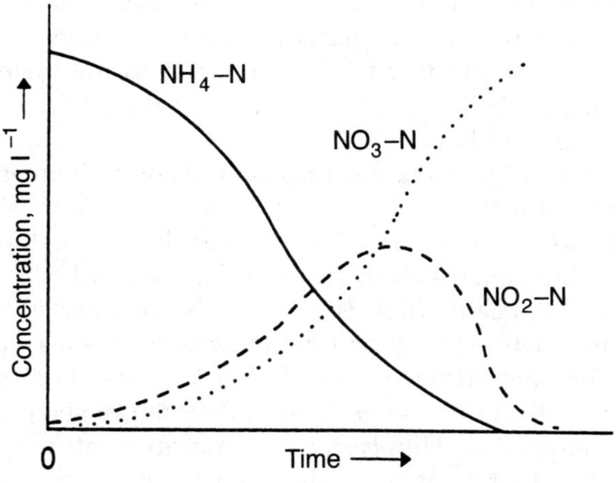

Figure 6.5 Nitrification curve showing the rise and fall of ammonia and nitrite and the rise of nitrate in closed aquaculture systems. (From Lawson, 1995.)

little effect on the concentration of ammonia in the wastewater. Consequently the ammonia concentration remains high downstream of the biofilter, whilst that of nitrite and nitrate is negligible (remember that these are the products of the consumption of ammonia by the micro-organisms).

Gradually, the *Nitrosomonas* community becomes established in response to the availability of ammonia as an energy source. As this population increases in size, so the ammonia concentration downstream of the biofilter progressively decreases. The product of this microbial activity is nitrite, and a concomitant rise in its concentration is seen.

The rising concentration of nitrite in response to the increasing size of the *Nitrosomonas* community in turn stimulates the development of the *Nitrobacter* community, which uses the nitrite as an energy source and so converts it to nitrate. Whilst nitrate is much less toxic to aquatic animals than ammonia or nitrite, high concentrations are nonetheless undesirable. The falling ammonia and nitrite concentrations and increasing nitrate levels shown in Figure 6.5 are as a result of the establishment of *Nitrosomonas* and *Nitrobacter* communities and the breakdown of ammonia to nitrite and then to nitrate. Control measures for levels of nutrients that rise above those recommended include regular water exchange or uptake by aquatic plants or algae. Additional recommendations for managing biofilter start-up are given in Box 6.2.

Biofilter seeding

There are some methods available for seeding biological filters so as to reduce the start-up time. These consist of bacterial preparations and/or preparations of inorganic nitrogen salts. They can be successful as long as certain basic facts are considered.

Irrespective of the quantity of nitrifying bacteria seeded in the filter, they will die if there is no ammonia for the population to consume.

Ammonia is often produced by putting fish in the system, but it is recommended that the initial source is through the addition of inorganic nitrogen salts to the filter.

If fish are used to provide the ammonia, then the biomass should be increased progressively. If this is not done there will inevitably be peaks of ammonia and nitrite in the effluent water due to the time lag in the response of the micro-organism populations, which may prove damaging to fish health (this is even if the biofilter is seeded with bacteria, as they will need some time to become established).

If using inorganic nitrogen salts, the best combination is to provide ammonia and nitrite salts so as to stimulate the development of both the *Nitrosomonas* and *Nitrobacter* communities at the same time (remember the double peak in the start-up performance curve). A solution of 12 mg N l^{-1} ammonium chloride and 12 mg N l^{-1} sodium nitrite has been used successfully in this respect (Lawson, 1995).

BOX 6.2
RECOMMENDATIONS FOR MANAGEMENT OF BIOFILTER START-UP

Time must be allowed for the microbial populations in biofilters to become established. This can take a number of weeks, depending on the size and function of the system, but there are some short-cuts. The following recommendations are intended to illustrate how the unit should be managed during the start-up period.

- Add wet medium from a mature system to seed it, or a commercial bacterial preparation.
- Condition the biofilter community with inorganic ammonia and nitrite compounds.
- Provide optimal conditions for microbial growth (sufficient O_2, constant temperature, adequate nutrition).
- Remember the influence of temperature on microbial activity, i.e. cool systems will mature more slowly than warmer ones.
- Allow plenty of time – plan ahead.
- For recycle systems, if the biofilter is to be in operation for only a short time (in a hatchery, for example), it is essential to consider whether a biofilter is the most appropriate method for nutrient and ammonia-N removal.
- Add the stock progressively, especially in recycle systems.

Wet filter medium from a mature biological filter can be used to seed a new biofilter very successfully. To avoid the transfer of disease between separate systems, only media from disease-free systems should be used.

6.1.4 Application of biofilters to aquaculture

Biofilters have two principal applications in aquaculture: as a central component of recycle systems, and for reconditioning wastewater prior to discharge.

(a) Recycle systems

Biofilters are an essential component of recycle aquaculture systems. Typically in recycle units, over 90% of the wastewater is recycled. Before returning to the culture tanks it must be reconditioned. This involves removing a substantial proportion of the dissolved nutrients, converting most of the ammonia to nitrate, removing the suspended solids, and reoxygenating the water. Biofilters have proved to be a reliable and cost-effective core of water reconditioning systems for the culture of fish, crustacea and shellfish.

(b) Wastewater treatment to meet discharge water quality standards

In many countries environmental regulations stipulate that aquaculture wastewater must meet certain water quality criteria before its discharge into natural waters is allowed. Meeting BOD and ammonia criteria are central components of such control systems, and biofilters can be used to condition wastewater to the requisite standard.

6.1.5 Biofilter design

Knowledge of the ecology and management of biofilters can be used to optimize their design.

(a) Biofilm kinetics

In an ideal world the knowledge obtained from the study of biofilm kinetics would be used to design biological treatment units. Biofilm kinetics is the study of the factors governing biochemical reactions within the biofilm. Because of certain limiting factors it is known that biochemical reactions (which control the conversion of ammonia to nitrate, for example) can only proceed at a certain rate and need a certain quantity of ingredients (e.g. ammonia and oxygen) to produce a certain quantity of product (e.g. nitrate). Thus we might be able to say what quantity of bacteria is needed to convert into nitrate the ammonia produced by each kilogram of fish.

However, biological treatment units have a complex ecology, with many different organisms and processes contributing to water treatment. Furthermore, these organisms do not exist passively – they continuously interact; for example, autotrophs may be excluded by heterotrophs where the influent concentration of dissolved organic compounds is high. Also, organisms change their own environment; for example, water quality will change as the water passes through a treatment unit, as a result of the uptake and processing of nutrients by the microbial community. As a consequence of this change in the environment, the community structure may change.

To model this change in order to produce a workable design formula for the biofilter is almost impossible. An alternative is to use a mass balance approach, based on the premise that what goes in must come out (or stay within the system). Lawson (1995) uses this approach in the design of aquaculture recycle systems and gives detailed examples. Hess (1981) provides a similar but more complex approach incorporating empirical data on nitrification characteristics of various biofilter configurations.

Whilst the mass balance approach is certainly very useful, it too has important limitations which must be recognized. Calculations used for designing a biofilter could be based on the most accurate data on respiration rate of the fish and the microbes but the results may still indicate 10%, 20% or even 50% error in size. It is not yet possible to size a real biofilter accurately, except by trial and error. The reason for this lies in the difficulties in predicting the waterflow patterns within a biofilter unit. These will vary not only with operating parameters (e.g. flow rate, aeration rate), but also with the shape and size of the unit. Short-circuiting, deadspace, turbulence and mixing rate are all relatively unpredictable factors. Theoretical predictions are likely to be closest to reality for some designs of rotating disc biofilter. Even using conventional designs, a system may underperform by 50% or more in some cases, compared with the theoretical predictions.

There are two practical alternatives to this approach. One is to buy a tried and tested off-the-shelf system for which the performance parameters are well known. The other is to engage in a research and development exercise, whereby a pilot system is built and tested over a period of at least 3 months (bearing in mind the start-up time common to all biofilters). Even then this must be expanded to a full-size system very carefully. Scaling up the dimensions of a pilot plant does not necessarily guarantee a proportionate increase in performance. Pilot plants should be designed as unitary systems which are scaled up by adding more units, rather than by increasing the dimensions.

Despite these difficulties the mass balance approach is still the only practical way of specifying the necessary biofilter capacity for a given culture system. This is the basic starting point in any design exercise.

(b) Types of biofilter

There are five basic configurations of biofilter: trickling or percolating; submerged; rotating-media; fluidized bed; and low-density media filters.

Trickling filter

Trickling filters comprise a column packed with filter media (Figure 6.6a). Wastewater is introduced at the top, by means of a rotating spraybar, a distribution grid or a tipping trough, and trickles down through the filter bed (Lawson, 1995). Thus, the biofilm is supplied with a flow of wastewater over its surface. As the bed is a three-phase system (air/water/solid), the water flowing through it is readily aerated. Trickling filters are less likely to suffer reductions in performance through oxygen deficiency than submerged filters.

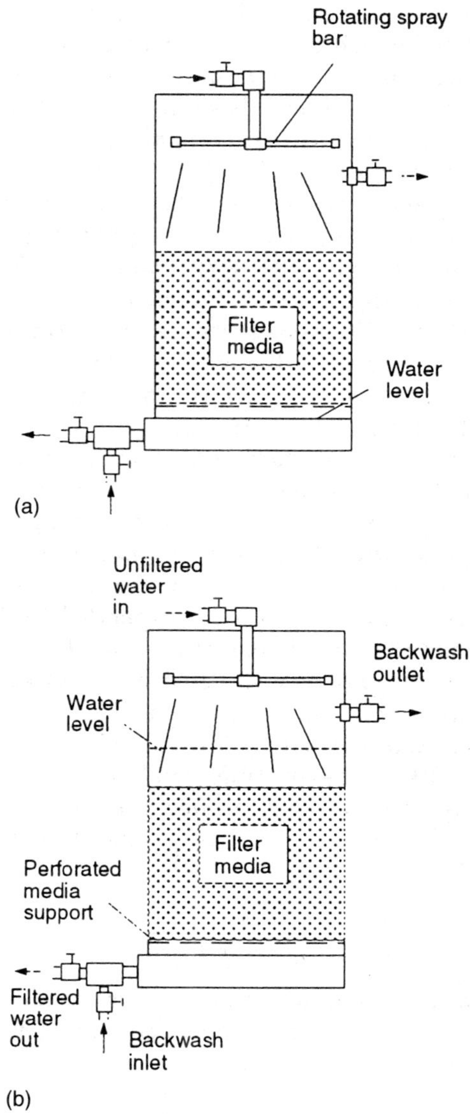

Figure 6.6 Biological wastewater treatment units: **(a)** trickling; **(b)** submerged. (From Lawson, 1995.)

Submerged filters

Here the filter medium is fully submerged in the wastewater (Figure 6.6b). The waterflow may be in a horizontal, upward or downward direction, depending on the location of inlet and outlet. Two-phase submerged filters (water and medium) are prone to oxygen shortages unless aerated, and short-circuiting can be a problem. Submerged filters

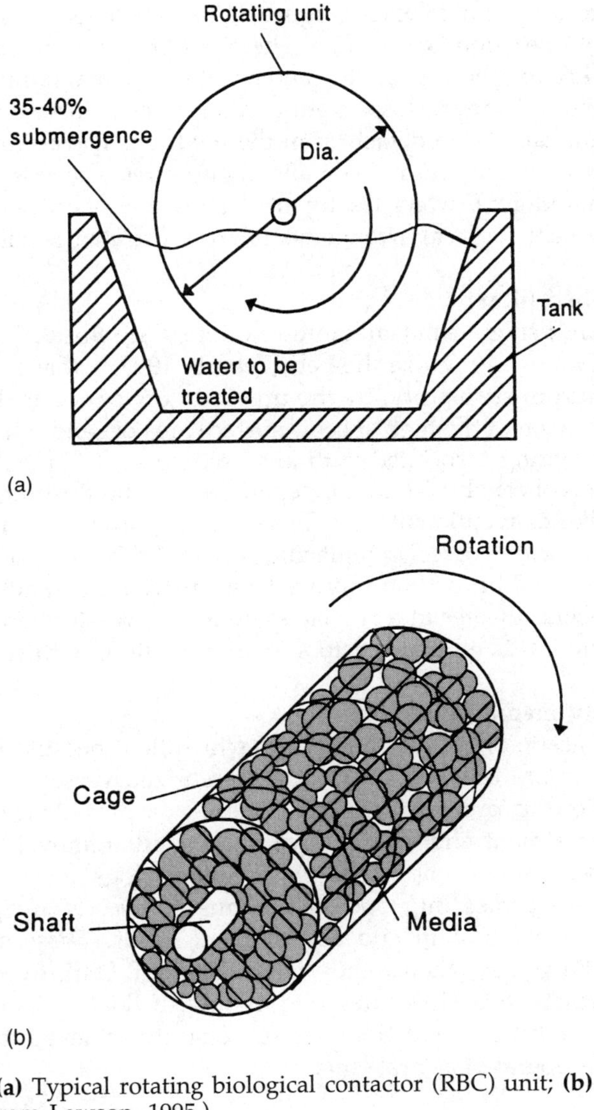

Figure 6.7 **(a)** Typical rotating biological contactor (RBC) unit; **(b)** rotating biodrum. (From Lawson, 1995.)

have a more stable environment for the development of microbial populations than trickling filters because of the mass of water they contain.

Rotating media filters

Typically, rotating biofilters use flat or corrugated discs as a filter medium (Figure 6.7a). Other media types can be used (Lawson, 1995), such as cylindrical medium packed into a cylindrical bundle, lamellar block-type medium fashioned into circular sections, or loose medium packed into a cylindrical cage (Figure 6.7b). Whatever the medium, the principle of operation is the same. The discs or drum are submerged to about 35–40% of their diameter and rotate at a maximum peripheral speed of about 0.3 m s^{-1}, to prevent excessive shear force on the biofilm which might cause it to detach from the medium. The alternation of the biofilm between the wastewater and air ensures adequate aeration of the biofilm and good water treatment. They are very efficient for wastewater treatment. Typical arrangements are illustrated in Figure 6.8.

Fluidized bed filters

Traditionally using a sand medium, the filters are flushed at high rates with wastewater to fluidize the bed (Tebbut, 1992) – that is, to hold the filter medium in suspension by the upward flow of water. Thus advantage can be taken of the high specific surface area of sand whilst avoiding the rapid clogging associated with low void space. In operation the bed increases in volume by 30–100%, depending on the flushing rate. These filters, unlike conventional sand filters, do not provide solids removal. They are not widely used in aquaculture, but Arbiv and van Rijn (1995) successfully used an anoxic fluidized bed reactor for denitrification of fish culture wastewater in a recycle system, and Westerman *et al.* (1996) used aerobic fluidized bed reactors for nitrification, with some success.

Low density media filters

These use plastic or polystyrene beads which float because of their low density. They are contained within a pressurized vessel, with a screen at the outflow to prevent their loss. The beads provide a medium for biological treatment and also achieve good solids removal by virtue of their relatively small void space. As the filter clogs a propeller-driven mixing device comes into operation. Thus the beads are periodically cleaned of excess biofilm and accumulated solids. These settle to the bottom of the vessel, whence they can be bled off. Hargrove *et al.* (1995) and Westerman *et al.* (1996) describe the use of floating bead filters for nitrification. Whilst nitrification was successfully achieved, both studies report some operational problems.

6.2 WETLAND TREATMENT SYSTEMS

Wetland treatment systems attempt to use the whole aquatic ecosystem for wastewater purification. These systems have been applied to sewage treatment, organic industrial effluents (especially those of food processing industries) and acidic mining wastewaters (Mitsch and Jorgensen, 1989; Cooper and Findlater, 1990; Etnier and Guterstam, 1991). The complexity of aquatic ecosystems and the limited knowledge of their dynamics has made the design of wetland treatment systems

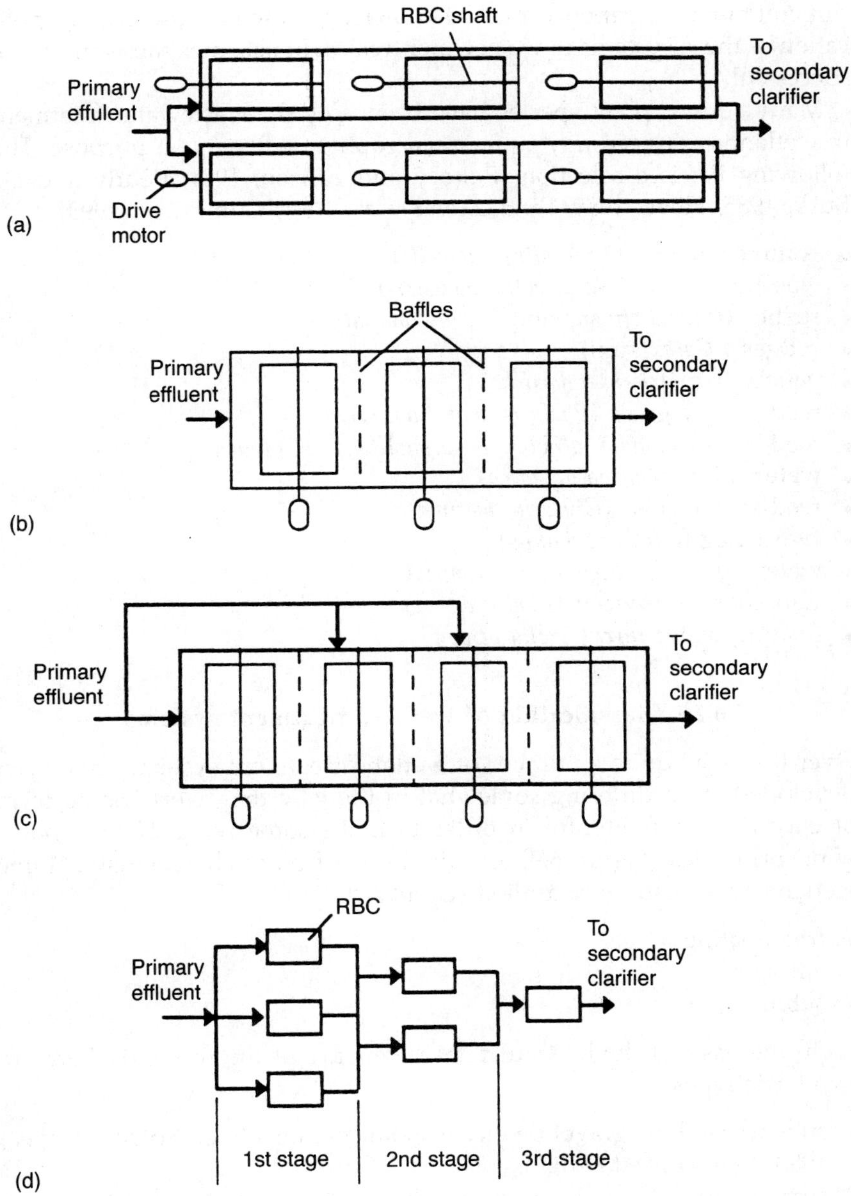

Figure 6.8 Typical RBC arrangements: **(a)** flow parallel to shaft; **(b)** flow perpendicular to shaft; **(c)** step feed; **(d)** tapered feed. (From Lawson, 1995.)

difficult but, as experience and the knowledge base increase, so the application of these systems to wastewater treatment becomes more and more successful.

Many aquatic plant species have been used for wastewater treatment in wetlands – indeed any aquatic macrophyte will serve a purpose. The following list is a selection (Rakocy and Allison, 1981; Reddy and De Busk, 1985; Hiley, 1990; Biddlestone *et al.*, 1991; Gumbricht, 1993):

- common reed (*Phragmites australis*)
- true bulrush (*Schoenoplectus lacustris*)
- rushes (*Juncus effusus* and *J. conglomeratus*)
- sedges (*Carex* spp.)
- yellow flag (*Iris pseudacorus*)
- reed canary grass (*Phalaris arundinacea*)
- reed maces (*Typha latifolia, T. augustifolia, T. glouca*)
- water mint (*Mentha aquatica*)
- reed sweet grass (*Glyceria maxima*)
- sweet flag (*Acorus calamus*)
- water hyacinth (*Eichornia crassipes*)
- Canadian pondweed (*Elodea densa*)
- floating water fern (*Azolla* spp.).

6.2.1 Classification of wetland treatment systems

Over the years quite a number of wetland treatment systems have been developed, each differing somewhat in the way they work. Some form of classification is helpful in order to make some sense of the underlying principles (Figure 6.9). On the basis of plant characteristics three wetland types can be identified (Gumbricht, 1993):

- free-floating
- emergent
- submergent.

On the basis of the hydraulic characteristics of the wetland, there are two basic types:

- subsurface flow: gravel-bed hydroponic systems (e.g. Wolverton *et al.*, 1983; May *et al.*, 1990);
- surface flow (with respect to substrate): planted wetland lagoons (Hiley, 1990).

In the following paragraphs the term wetland treatment systems refers to planted wetland lagoons, which may have floating, emergent and submergent vegetal components. Gravel-bed hydroponic systems are more complex in design and it is questionable whether they have any advantages over wetland lagoons for aquaculture wastewater treatment.

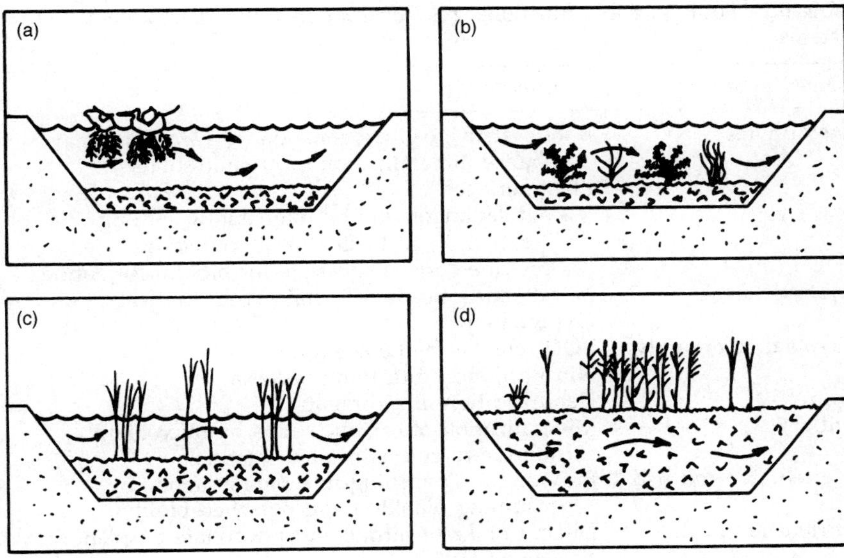

Figure 6.9 Cross-sections of different types of constructed aquatic treatment systems: **(a)** free-floating macrophyte system; **(b)** submersed macrophyte treatment system; **(c)** constructed wetland with emergent plants and free water surface; **(d)** constructed wetland with emergent plants and subsurface horizontal water flow. (Redrawn after Gumbricht, 1993.)

6.2.2 Special nature of aquaculture wastewaters

Wetland treatment systems have primarily been applied to the purification of wastewaters with high BOD loading – high, that is, in comparison with aquaculture wastewaters. Tertiary sewage effluent typically has a BOD in the region of 100–200 mg l^{-1}. In aquaculture wastewaters the BOD is more likely to be 8–10 mg l^{-1}. Furthermore, aquaculture wastewaters are normally high volume. It is not usually practical to allow batch treatment of the wastewater of several days' duration, as is common practice with the more conventional high concentration/low volume wastewaters. The two basic problems facing the application of wetland treatment systems to aquaculture are:

- low strength/high volume wastewater;
- short residence times.

The implications of these factors for wetland treatment system design are discussed later in this chapter.

Table 6.1 Some possible functional relationships in wetland treatment systems

Component/process	*Functional role*
Macrophytes	Root systems, e.g. oxygenation by *Phragmites* may enhance mineralization of organic solids Plant surfaces: • aid sedimentation by intercepting settling particles and facilitating flocculation • provide surface substrate for biofilm formation
Epiphytic algae	Nitrate and phosphate uptake from the water column Oxygenation
Microbial community	BOD removal by heterotrophs Nitrification by nitrifying bacteria Denitrification in sediments
Benthos	Mineralization of organic solids by microbial community and associated infauna.
Invertebrates and fish	Provide final polishing in separate lagoon by consuming plankton and detached biofilm.
Nutrient removal	Direct uptake of nitrate and phosphate by plant surfaces (?) Uptake of nutrients from the water column through adventitious root systems Uptake by plants from (aerobic) sediments Plant constitute sinks for N and P, allowing removal from system by cropping
Sedimentation	Plant matrix modifies water flow patterns, so facilitating sedimentation See also Macrophytes (plant surfaces)
Oxygenation	Plant photosynthetic activity generates oxygen, which is released into the water
Inorganic sediments	Long-term phosphorus sink

6.2.3 Functional relationships in wetland treatment systems

The functional relations of the wetland treatment systems can be inferred from the knowledge of natural aquatic ecosystems but it is important to recognize that, in most cases, there is insufficient data even to confirm the functional relationships suggested here, let alone to quantify them for design purposes. Wetland treatment systems may include the components and processes shown in Table 6.1.

The following accounts illustrate some of the complex interrelations between the functional components of wetland treatment systems. It is just possible to discern some fundamental principles that point the way for future research and development, and these are discussed later.

(a) Phragmites in wetland treatment systems

In systems that use emergent plants such as the common reed, the submersed roots and stems have several functions. They may take up nutrients directly, provide a surface for microbial communities to establish, and aid the sedimentation of solids. In addition to these functions, *Phragmites* has an internal oxygen transport mechanism that is thought to be important in the dynamics of reedbed systems (Armstrong *et al.*, 1992): oxygen can be passed through the stems to the rhizome and to the root hairs, creating an oxygen-rich environment around the root zone that is suitable for the growth of aerobic micro-organisms in an otherwise anaerobic environment. The presence of both aerobic and anaerobic conditions around the root system may enable the processes of both nitrification and denitrification to occur. The microbial population is also involved in the mineralization of organic matter, thus contributing to lower BOD. Reddy *et al.* (1989) found similar processes in respect of a range of aquatic macrophytes used for wastewater treatment.

(b) Nitrogen dynamics in wetland treatment systems

Nitrogen is taken up directly in inorganic forms, i.e. ammonium (NH_4^-) and nitrate (NO_3^-), by aquatic macrophytes and algae, but plant stems and leaves suppress algal growth through shading (they also reduce surface and subsurface water movements, thereby reducing turbidity). The processes of nitrification and denitrification are carried out by specific aerobic and anaerobic bacteria. In dense macrophyte stands the plants provide a considerable surface area for microbial attachment. The transformation of nitrate to nitrogen usually takes place in anaerobic conditions in the sediment matrix. In this way nitrogen compounds are removed from the system. However, the floating fern *Azolla* can form a symbiotic relationship with a blue green alga, *Anabaena azollae*, which has the ability to fix atmospheric nitrogen (Kitoh *et al.*, 1993), bringing nitrogen back into the system when it is limiting. This process is inhibited by higher concentrations of ammonium and the plant is also sensitive to high nitrite concentrations.

6.2.4 Aquaculture and wetland treatment systems

The application of wetland treatment systems for the treatment of aquaculture wastewaters is new and often experimental. Case studies 6.1, 6.2 and 6.3 describe some of the recent trials that have been undertaken to investigate the potential and practicalities for the use of wetland treatment systems for aquaculture wastewater.

CASE 6.1: CATFISH IN THE USA (Schwartz and Boyd, 1995)

In Alabama, wastewater from ponds containing commercially reared channel catfish (*Ictalurus punctatus*) was passed through a specially constructed wetland system (Figure 6.10). It consisted of two wetlands,

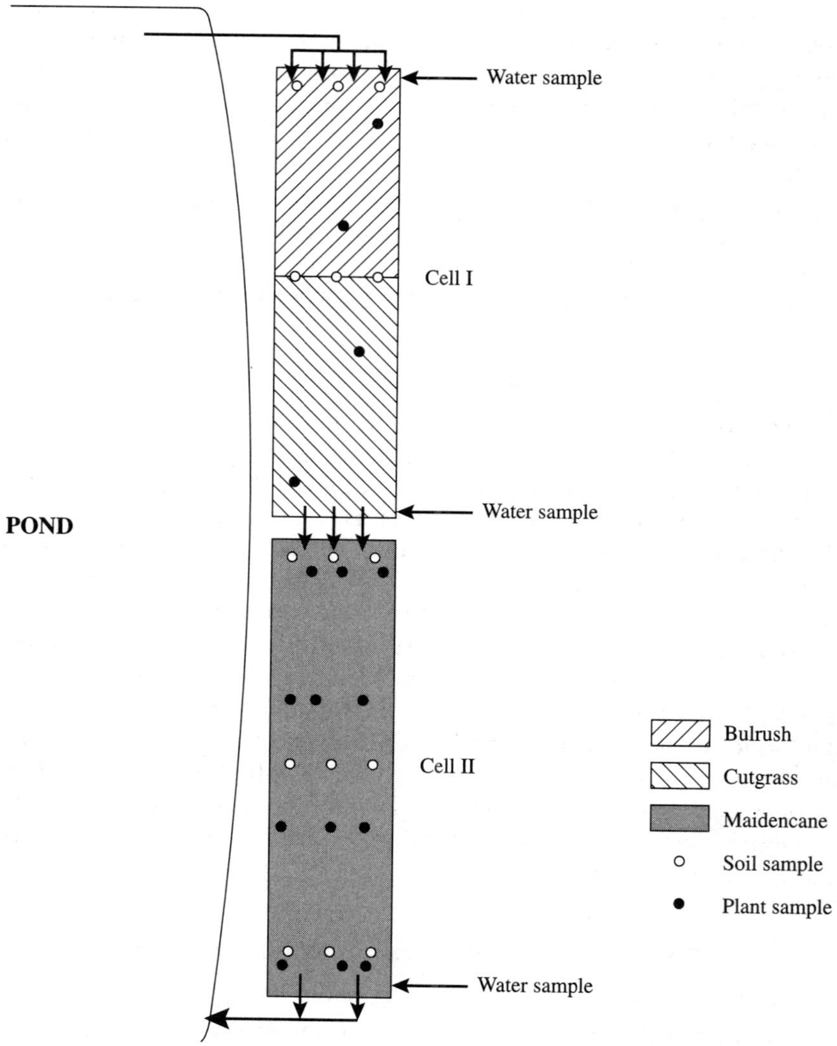

Figure 6.10 Location of constructed wetland-to-pond water delivery scheme, and location of plant and soil collection sites. Water was pumped from a 6.9 ha channel catfish pond into Cell I, which contained *Scirpus californicus* and *Zizaniopsis miliacea*; it was allowed to flow by gravity into Cell II, which contained primarily *Panicum hemitomon*, and then exited from Cell II to drain back into the pond. Both cells measured 84 × 14 m. (From Schwartz and Boyd, 1995.)

arranged in series. The first was planted with Californian bulrush (*Scirpus californicus*) and giant cutgrass (*Zizaniopsis miliacea*), and the second with Halifax maidencane (*Panicum hemitomon*). All three are emergent plants.

This system removed on average 36.9% of BOD_5, 75% of suspended solids, 52.7% of nitrate and 43.9% of nitrite overall. Acceptable removal rates were recorded even during the colder months of the year, when plant growth was dormant.

The removal efficiency was affected by the hydraulic loading rate (HLR)(l m^{-2} per day) and the hydraulic residence time (HRT). Hydraulic residence time was calculated as:

$$\text{HRT} = AD(1 - P)/Q$$

where A is the area (m^2), D is the depth (m), P is the proportion of water volume occupied by plants (dimensionless) and Q = (inflow + outflow)/2 (m^{-3} per day) .

In order to provide the optimum residence time for a specific volume of water, the area of the treatment system required could be calculated by the equation:

$$\text{Area} = V/(\text{HLR})T_d(10^{-3})$$

where V is the pond volume (m^3) and T_d is the time of pond draining (days).

Thus a 1 ha fish pond with a flow of 80 l m^{-2}, a depth of 1.5 m and a draining time of 7 days would require a treatment area of 2.5 ha. Note that the treatment area is 2.5 times the size of the fish pond. This places a significant constraint on the application of wetland treatment systems of this type to commercial aquaculture situations under present regulatory and economic circumstances.

CASE 6.2: ELODEA AND ORNAMENTAL FISH PRODUCTION (Ng et al., *1990)*

In Singapore, at a tropical ornamental fish farm, the submergent aquatic macrophyte *Elodea densa* was assessed for its ability to remove nutrients and recondition fish pond wastewater. Trials using a small (11.4 m^3) tank planted with *Elodea* achieved nitrogen (ammonia, nitrite and nitrate) removal of approximately 25%. The plants also reduced turbidity, and prevented the growth of phytoplankton by shading. In fact the planted system was more effective in reducing turbidity than a fine gravel filtration plant. Reduced removal rates were recorded as plants were periodically harvested during the trial, but the recovery was fairly rapid (2–3 days). It was concluded that this system would serve the dual purpose of water renovation and the production of a saleable product.

CASE 6.3: FISH CULTURE AND WATERCRESS *(Redding* et al., *1997)*

In the UK, in a trial using a recirculation system for the culture of tilapia, the rates of uptake of nutrients by three species of aquatic plant were compared. Plant treatment units contained monocultures of watercress *Rorrippa nasturtium-aquaticum* and two other species of aquatic macrophyte, *Azolla filiculoides* and *Elodea nuttalli.* All three systems consistently reduced the levels of ammonia-N, nitrate-N and phosphorous in effluent from the culture tanks. It was concluded that the residence time was an important factor in the effective removal of nutrients and this may be difficult to provide in small experimental units. However, the system using *R. nasturtium* proved to be the most effective at reducing nutrient levels. Figure 6.11 illustrates the removal efficiency (expressed as percentage change in concentration) of the three different systems compared in this study (against an unplanted control). The *R. nasturtium* system achieved over 40% reduction in ammonia concentration through the system, up to 20% reduction in nitrate concentration and up to 20% phosphate reduction.

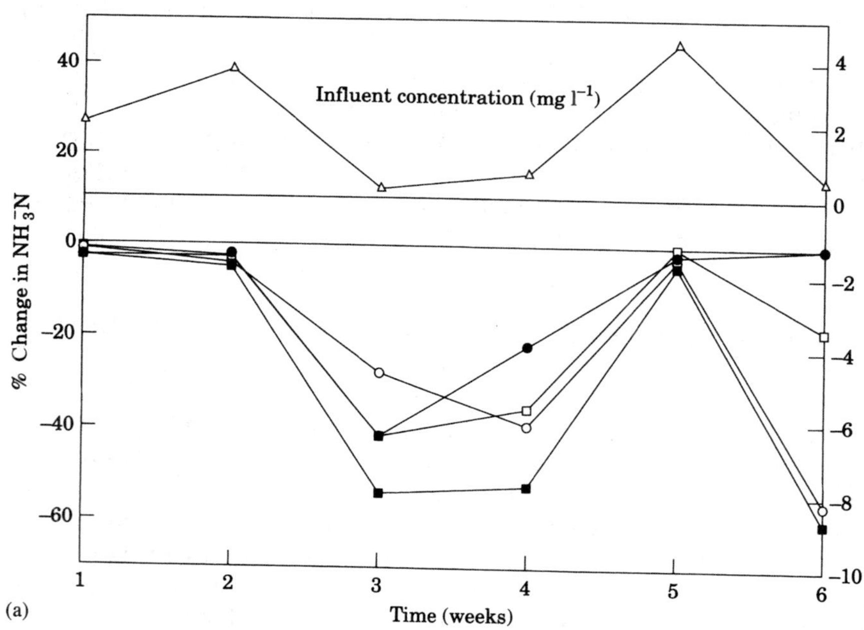

Figure 6.11 Weekly mean percentage changes in concentration of **(a)** ammonia-N, **(b)** nitrate-N and **(c)** phosphate: ● free-floating; ○ submersed; ■ emergent; □ control. The emergent plant is *R. nasturtium*. (From Redding *et al.*, 1997.)

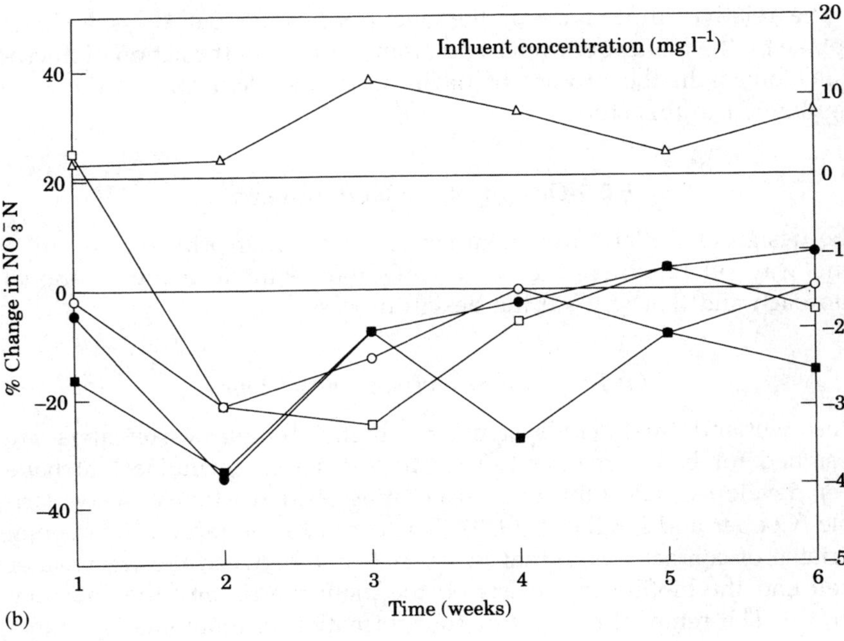

Figure 6.11(b)

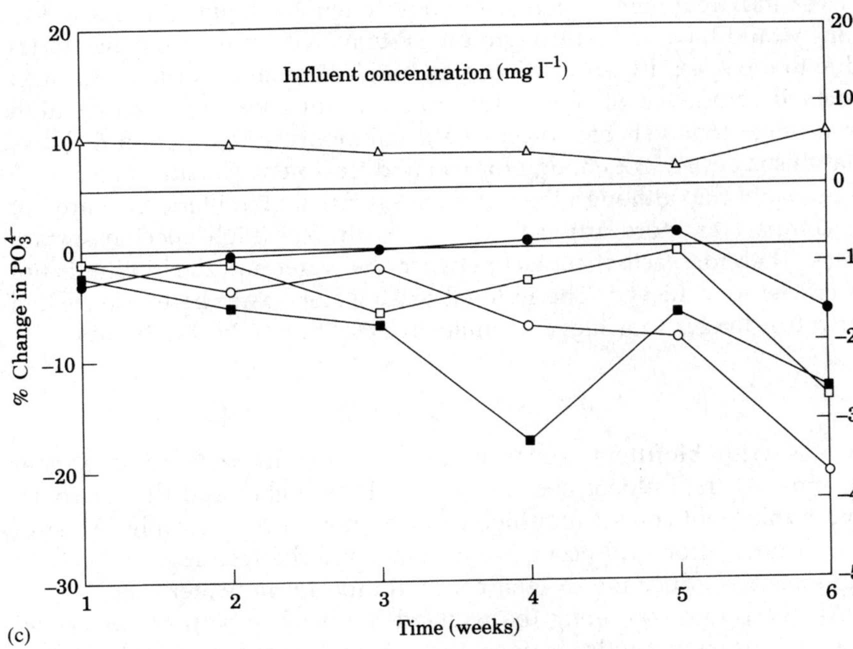

Figure 6.11(c)

The relative importance of the two purification pathways – direct uptake by the plants of nitrates and ammonium, and the action of microbial biomass in the process of nitrification and denitrification – were highlighted in this study.

6.2.5 Design of wetland systems

The design of wetland treatment systems from first principles is still a long way off but there are some clues that point to a more rational approach and that will aid the design process.

(a) Residence time versus contact time

Most wetland treatment systems for high strength wastewaters are designed for batch treatment. The various design formulae that have been developed reflect this by incorporating residence time as a key variable (Cooper and Findlater, 1990). This obscures the more fundamental variable of contact time – that is, contact time between the wastewater itself and the biofilm that exists on the plant matrix and the substrate surface. The removal of BOD or transformation of ammonia is a function of contact time, and is expressed as percentage removal rate. This terminology is common in the design of biofilters, which are throughflow systems.

Wetland treatment systems for most intensive aquaculture applications would have to be throughflow systems with short residence times (due to the necessity for continous and high volume waterflow). Contact times in biofilters of short duration are sufficient to achieve high percentage removal rates of BOD and ammonia, but experience shows that this is clearly not enough for wetland treatment systems. The reason is probably that although the latter are essentially biofilters, they are not as compact as more artificial systems with very high specific surface areas. Therefore, actual contact between the water and the biofilm is not as efficiently achieved. The influent wastewater has to be in the system for much longer to achieve a similar contact time with the biofilm.

(b) Zonation

Just as with biofilters, zonation can develop in wetland treatment systems. At the influent end the BOD will be higher and the microbial community will consist of a higher proportion of heterotrophs. Most of the sedimentation will occur in this zone, and the resulting BOD could significantly reduce the oxygen concentration in the water.

As one progresses along the wetland the balance between BOD and nitrogenous compounds changes, and this will be reflected in the microbial community. In wetlands used for high strength wastewaters, only very poor rates of nitrification have been reported (Cooper and Findlater, 1990). This is probably because the effluent BOD is relatively high and still favouring heterotrophic microbial populations. In aquacultural systems the influent BOD is already relatively low, and acceptable rates of nitrification are achievable. It is important to provide a long enough wetland to allow this zonation to develop, and something approaching 100 m in length has been suggested as appropriate (Hiley, 1990).

(c) Oxygenation

Submersed macrophytes release oxygen into the water during daylight hours, as a result of photosynthetic activity. At night, they consume oxygen and so effluent concentrations of dissolved oxygen may be below requisite levels. In systems based on floating macrophytes, oxygen levels have been found to be low – presumably due to the small underwater plant biomass and a reduced amount of surface exchange of oxygen.

6.2.6 Wetlands versus biofilters

The potential advantages of wetland treatment systems over biofilters include cost considerations and that wetland treatment systems are likely to achieve a more complete reconditioning of the water than biofilters, by providing effective solids removal, mineralizing nitrogen to its gaseous form and oxygenating the water, for example. There is also the potential for crop production from the wetlands (see below) to offset the capital and operating costs. Finally, their wildlife value must not be overlooked.

(a) Crops

The cropping of biomass in wetland treatment systems is recommended in order to remove C, N and P from the system. Some uses for crops produced in this way are:

- *Elodea* spp. as garden pond plants;
- watercress as a food crop;
- fodder crops for livestock (e.g. *Phalaris* sp. grass);
- willow for crafts or fuel;
- rushes and reeds as craft materials;
- plant biomass for biogas production and compost.

(b) Constraints

In addition to the large area required for effective reduction of pollutants, the climate may be a significant constraint to the use of wetlands at higher latitudes (Lakshman, 1979). In temperate climates wetland systems will probably provide adequate treatment in the warmer months, but are unlikely to cope with loadings typical of an intensive fish farm during winter. However, the winter months are often less critical with regard to the environmental impact of wastewater discharges, because of the increased river flows at that time.

6.2.7 Bacterial augmentation

In the past, bacterial augmentation (specially prepared bacterial suspensions) has been used in aquaculture (e.g. Boyd *et al.*, 1984) to improve pond water quality or to aid digestion of organic sediments, but without sufficient understanding of microbial ecology; the practice has, therefore, seen mixed success. There is now a better understanding of microbial ecology and greater success is being seen in the management of microbial communities in aquaculture ponds.

The most important aspect of microbial ecology in this regard is competitive exclusion – the exclusion of one species by another that has some competitive advantage over the first (Moriarty and Body, 1995). There are two factors at work here: chance, and physiological capability (including growth rate, ability to produce antibiotics, and resistance to adverse physicochemical factors).

It is no surprise that physiological capability is a dominant factor in competitive exclusion but the role of chance may also be very important, at least in the short-term. There are a great many strains of bacteria, many of which may have the right physiological capabilities to exploit a particular set of environmental conditions. Chance may be the deciding factor with regard to what determines the dominance of one species over another with very similar capabilities. The environment, at the scale of a bacterium, is highly variable, both in space and in time. Being in the right place at the right time may tip the balance in the favour of one species at the expense of another.

Management of microbial populations in aquaculture ponds has two elements:

- seeding with appropriate beneficial bacteria to maximize the chance that the right bacteria is around at the right time;
- manipulating the physicochemical conditions in the pond so as to favour beneficial bacteria and exclude pathogens.

To illustrate this last point, *Vibrio* spp. survive unfavourable conditions by attaching to algal cells (Moriarty and Body, 1995). If there is an algal bloom in a pond of those species that can be utilized in this way by pathogenic *Vibrio*, then a bloom of pathogenic *Vibrio* following the algal bloom can be expected. During the initial stages of an algal bloom heterotrophic bacterial numbers are limited, as the release of organic molecules by the algae is small. The ability of *Vibrio* to survive this stage in relatively high numbers by attachment to algal cells would allow them to colonize the water column rapidly as the algal population decays.

Theoretically, bacterial augmentation has considerable potential as a measure for reducing the dependence on antimicrobials for the treatment of disease in aquaculture (Chapter 2) and for enhancing the management of water and sediment quality in aquaculture ponds.

SUMMARY

- Biological wastewater treatment systems are suitable for the treatment of aquaculture wastewaters.
- A study of aquatic ecology provides the basis for understanding their operation, and for their design.
- Biological filter systems include:
  - trickling filters
  - submerged filters
  - rotating media filters
  - fluidized bed filters
  - low-density media filters.
- Wetland treatment systems, utilizing a wider range of ecosystem processes than conventional biological filters, may have applications for aquaculture.
- Bacterial augmentation and the manipulation of the microbial ecology of aquatic systems to achieve wastewater treatment is an emerging field, and has potential applications in aquaculture.

CHAPTER 7

Environmental management systems for aquaculture

In the 1980s, total quality management (TQM) became perhaps the leading business ethos. For the uninitiated, TQM is about getting it right every time – the benefits being increased customer satisfaction, and thus loyalty; and less time and money spent rectifying errors, and thus a more efficient business.

At the heart of TQM is an appropriate management system, one in which every process or procedure has a reference manual (a few lines or many pages, depending on its complexity) (Jackson and Ashton, 1993; Waller *et al.*, 1993). Such a manual should contain instructions and checks to make sure that the work is properly done, or that errors can be rectified before another part of the business or the customer is affected. This is a cyclical process (Figure 7.1). TQM is standardized, and businesses are accredited as having installed an appropriate management system under the international standard ISO 9000 series (BS5750 in the UK; EN29000 series in the EU).

A more recent development is the Environmental Management Standard BS7750, launched in 1992 by the British Standards Institute (BSI). Environmental management systems (EMS), which can be operated as an integral part of a TQM system, are designed to improve the environmental performance of businesses, i.e. to reduce the environmental impact of their activities. Whilst the key goal of TQM is zero defects, the key goal of an environmental management system is zero negative environmental impact. Logically, this is not achievable, since all human activities have some environmental impact – even cleaning up waste has an impact on the environment. However, aiming for the unattainable will ensure continuous improvement of environmental performance, and this is a key element of BS7750.

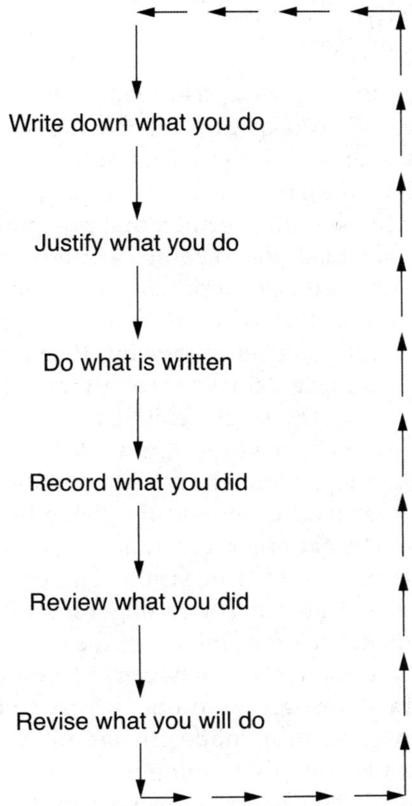

Figure 7.1 BS5750: a continuous cycle of improvement. (From Welford and Gouldson, 1993.)

CASE 7.1: IDAS TROUT LTD, IRELAND
(O'Laoire and Welford, 1994)

The management of IDAS Trout Ltd was prompted by a regional initiative to implement an environmental management system. The company's market research had shown that its location in the Irish Republic, allied to increasing customer preference for environmentally friendly products, could give it a competitive advantage in European markets. The EMS would give its customers an assurance of the environmental quality of its products.

At the time, the business consisted of a hatchery and on-growing facilities, and also fish filleting and smoking operations. The environmental issues facing the business were:

- husbandry and management systems;
- resource conflict issues;

- waste management and disposal;
- health, safety and hygiene.

As was considered in Chapter 2, fish feed is the major input to most intensive fish farms and feed quality can have a significant impact on fish growth rate, FCR and pollution. This link was emphasized in the environmental review undertaken as the first step in implementing the company's EMS: the system specified that low pollution diets should be used, and also specified the feeding regimes, as the timing and methodology of feeding are also important in reducing feed wastage.

A related problem was that of predation by herons. The farm had learned to live with a certain level of loss, but the environmental review showed that herons did more damage through the stress that their presence caused to the rest of the stock. This led to reduced appetite (and potentially more pollution as feed may be wasted) and increased risk of disease. Consequently, steps were taken to protect the stock from herons – a task that had not seemed economically viable before.

Another issue raised by the environmental review was the use of pharmaceutical chemicals in disease prevention. The problem of increasing resistance to pharmaceuticals by fish pathogens resulting from prophylactic use was identified. This recognition of the dangers of prophylactic treatment stimulated a change in management emphasis from that of maintaining a certain stocking rate to one of maintaining stock survivability. Whilst this may seem a subtle change in outlook, it can result in more environmentally friendly farming practices. It leads to questions such as: 'At what stock density must fish be kept, in order to ensure 90% survival?' rather than: 'Fish must be kept at a high stock density: what chemical treatments should be used in order to ensure 90% survival under these conditions?'

It can be seen immediately that this approach may lead to a variety of benefits, such as reduced chemical usage and fish stress, and improved fish health, FCR and flesh quality. All of these are associated with reduced stock density. However, reducing stock density is not commonly perceived as a positive management measure, i.e. one leading to improved product quality (which nowadays often includes an environmental component in the mind of the discerning customer) and competitive advantage. Instead it is traditionally identified with reduced efficiency (based on space and water usage) and reduced income. As is shown, the experience of IDAS Trout suggests that this attitude is becoming outdated.

The implementation of the EMS also brought about a cultural change in the workforce. Quality became central to the role of everyone in the organization – the success of the business, their jobs and their future depended on good environmental practices. Without this understanding and commitment from the workforce as a whole, the success of the EMS at IDAS Trout would probably have been limited.

Using its new environmental management system as a marketing tool, IDAS Trout Ltd saw an immediate increase in orders in its main European markets.

7.1 ENVIRONMENTAL MANAGEMENT SYSTEMS

It is not the place of this book to detail management systems, their design and their implementation (further reading is indicated in the text). A quality management system is one in which operating procedures for all aspects of the business are documented in a systematic way. They are there to assist staff in their work so as to ensure that the business runs smoothly and serves its customers well. However, a quality management system is not just a set of procedures. It must recognize that mistakes are made and that changes are needed, and so it must include systems for problem identification and for investigation and rectification. In the face of changing customer needs and product specifications this is an ongoing process. It is cyclical; it is a system.

The key elements of the environmental management system (Welford and Gouldson, 1993) are:

- the environmental review;
- the environmental policy;
- system design and implementation;
- the environmental audit.

7.1.1 The environmental review

Before an appropriate environmental management system can be designed for any business, it is first necessary to establish how the business's activities impact on the environment and to what extent. A list of processes should be compiled, and against each item an indicator or estimate of its impact on the environment should be entered. For an aquaculture business a sample from the list may be similar to that in Table 7.1 (impact scores can be calculated in the same way as for EIA; see Chapter 4). This process can also be applied to environmental policies and development projects (World Bank, 1991).

The pro forma in Table 7.1 is just a starting point. An analysis of consumption, production, discharges (including atmospheric and to mains sewer, and refuse collection), disposals and stocks should enable the identification of undetected waste, using the adage 'what goes in must come out', i.e. inputs should equal stocks plus outputs. If they do not, there must be a check for undetected sources of waste.

An assessment of the impact on the environment is a key part of the review. It is important to identify the mechanism of impact, in order to

Table 7.1 A pro forma for environmental review for an aquaculture business

Processes	*Energy consumption*	*Organic waste*	*Chemical waste*	*Non-renewable resource*	*Renewable resource*	*Hazardous chemicals*	*Stress, injury or death to wildlife*	*Habitat destruction*	*Noise*	*Visual impact*	*Risks*	*Impact score*
Feeding fish												
Transport of fish to market												
Stock health management												
Water abstraction												
Water discharge												
Tank/pond cleaning												
Operation of filtration equipment												
Anti-predator measures												
Fish processing												

Note: This should not be taken as a complete pro forma, but just as an illustrative example. Any business will have many more aspects of their activities which may have environmental significance, and all should be evaluated.

implement changes. Records should be kept concerning all legislation and regulations affecting the business; for example, abstraction and discharge consents, health and safety regulations, planning consent conditions, etc. These will form the basis of minimum standards for environmental performance.

Finally, a thorough review of existing management practices should be undertaken, with environmental performance in mind. Questions that should be asked might include the following. Is there a recognizable management system in place already? Does it include any environmental management objectives? If so, are procedures for monitoring, recording and reviewing environmental performance adequate? Are there any gaps in the system which may lead to poor environmental performance – does everyone know what to do in the event of an accident?

7.1.2 The environmental policy

In large organizations, it is vital to have a clear environmental policy that everyone within the organization understands and accepts. Without it people will be working towards different and perhaps contradictory objectives, and the development of an effective environmental management system would be impossible. The implementation of an EMS must start with a clear statement of environmental policy (Welford and Gouldson, 1993). Everything else – the design, monitoring and review of the system – can then be related specifically to that policy.

In a small business (such as many aquaculture businesses), communication is much simpler because there are few management layers and relatively few staff in total, but a clear environmental policy is no less important. The very process of writing an environmental policy is itself a valuable task. It will help the management and staff (everyone should be involved in the process) to focus on the most important environmental issues facing the business, and in what areas the most improvement could be made and what can be achieved realistically. Without going through this policy-making stage, it would be difficult to know where to start in designing an environmental management system. As detailed in later sections, there are also marketing benefits to be had from having a clear environmental policy.

7.1.3 System design and implementation

The design of management systems needs to be a painstaking process – small errors or omissions could have serious consequences. This is particularly important where a number of people or parts of an organization are involved in the procedure. For this reason it is important that everyone in the business be involved at some level in the design of the management system. It is no good the manager or a consultant writing a procedures manual for a task he or she has never performed. The people actually doing the work will be able to spot problems and omissions better than anyone else. The role of managers or consultants should be as facilitators of this process.

An environmental management system for a typical aquaculture business would address at least the following aspects of the business (adapted from Welford and Gouldson, 1993, and Welford, 1994):

- Short-term objectives for environmental impact reduction (related to long-term policy) and specific actions designed to meet these objectives.
- Designation of responsibilities under the EMS – for example, who monitors overall performance, who records the amount of feed used every day, who is responsible for chemical safety, who writes quarterly reports, etc.
- What staff training is to be undertaken in this period and how that is related to environmental performance. Is this the best use of training resources with regard to meeting EMS objectives?
- How the EMS is to be monitored and reviewed. Specific procedures and timetables should be laid out at this stage.

7.1.4 The environmental audit

The environmental auditing procedure (Aldridge, 1994) compares the business's performance with its environmental policy objectives and legislative obligations. It should be undertaken at specified intervals. A vital component is the review of policy in the light of past progress towards the objectives, and recommendations for improvements during the next period.

7.2 ENVIRONMENTAL MANAGEMENT SYSTEMS AND POLLUTION CONTROL

This section addresses how an EMS can enhance efforts to reduce pollution. Environmental management systems have a number of elements that contribute to this overall objective (Welford and Gouldson, 1993: Waller *et al.*, 1993):

- The **Environmental Policy Statement** clearly sets out long-term goals, and specific targets for improved environmental performance as a central element of business development (Box 7.1). Pollution is the main factor in environmental performance in aquaculture businesses.
- The **Procedures Manual** sets out what should be done and how it should be done for all activities in the business. In an environmental management system, procedures are designed to minimize pollution

Table 7.2 Pro forma for an environmental monitoring system

Variables	*Recording method*	*Recording frequency*
Environmental		
Water flow in and out	Instream device	Continuously or daily
SS	Filter, dry, then weigh	Weekly and during flushing
BOD	Meter or kit or contract, 24 h integrated samples	Monthly
NH_4	Test kit (HACH type)*	Daily
NO_2	Test kit (HACH type)*	Daily
NO_3	Test kit (HACH type)*	Daily
pH	Meter	Daily
Temperature	Thermometer	Daily
Salinity	Meter	Daily
Benthic invertebrates	Contract to specialists	Annually
DO	Meter	Daily
Process		
FCR	Record sheets/sample weighings	Monthly
Sludge volume	As appropriate	As appropriate
Feed specification	Record sheets	Feeding time

*Ormaza-Gonzalez and Vilalba-Flor (1994)

> BOX 7.1
> A SAMPLE ENVIRONMENTAL POLICY STATEMENT (EPS)
>
> The Global Aquaculture Company recognizes the vital role a healthy environment has played and will continue to play in the success of this business. We are therefore committed to minimizing waste in all our activities, and to achieving continual improvements in environmental performance. We believe that this is in the long-term interests of our company, our customers and future generations.
>
> Note that the EPS:
>
> - states the importance of maintaining environmental quality;
> - recognizes the link between environmental health and business success;
> - commits the company to improving environmental performance.

and to ensure that mistakes which could cause pollution or increase waste are not made.

- An **Environmental Monitoring System** requires the systematic recording of data, and the systematic analysis of that data in order to review environmental performance. This provides essential feedback

BOX 7.2
MONITORING PERFORMANCE

Monitoring performance is an essential component of any environmental management system. It provides vital feedback to the whole process, and enables the review of performance.

Table 7.2 gives some examples of process and environmental variables which should be measured and recorded on a regular basis. Variables should be chosen that are indicative of those aspects of environmental performance that have been targeted for action.

Correct recording procedures require the following.

- A record sheet is kept for each variable.
- Summary sheets for pre-defined periods are compiled from analysis of the record sheets (calculation of means, ranges; identification of data that falls outside target values).
- Performance is reviewed regularly. Reports should be compiled summarizing records, and interpreting performance by reference to the company's environmental performance targets. Specific problems should be identified and remedial measures proposed. For example, were discharge water quality standards met? If not, why not? How will it be ensured that this does not happen again?

The review reports should do more than identify problems. Recognizing activities that have been particularly successful is equally important, raising such questions as 'Why?', and 'What lessons can be learned?' The report may also consider and assess the potential benefits of new equipment or management procedures. This analysis will lead to the recommendation of new environmental performance targets and recommendations for their attainment.

to the environmental policy which helps to prioritize activities and set targets for improvement. Table 7.2 gives an indication of what environmental and process parameters should be measured, and how and when. This is not a list everyone should use, since it is important to tailor monitoring programmes to individual needs and situations, but it should give some idea of where to start in the design of such a programme (see also Box 7.2).

7.3 ENVIRONMENTAL MANAGEMENT SYSTEMS AND ORGANIZATIONAL ETHICS

The environmental policy statement should encourage a whole new attitude to all aspects of the business. Some aspects of business management of which this is particularly true are discussed in the remainder of the chapter.

7.3.1 Accident procedures

In the event of an accident or emergency – such as a chemical spill into a river, or a major power failure – it is important that everyone knows their responsibilities. When facing prosecution for environmental damage (a substantial fish kill, for example) it is not sufficient to offer the excuse that you thought so-and-so had already called the emergency services or relevant environmental protection agency, or that you did not know where to get the key for the store-room where the emergency equipment was kept. Everyone should know what to do and what everybody else is doing as well. Therefore, procedures should be recorded in an understandable and accessible form, and regular training should be given so that procedures are fully understood.

7.3.2 Training

Staff training is a vital aspect of quality management. It is through attending to the development of skills amongst the workforce (management as well) that quality will be maintained and improved (Jackson and Ashton, 1993). This holds true for environmental quality management as well. It is essential that staff understand the need for an environmental management system, and how they fit into it. They must also be able to carry out their duties correctly and reliably under the system, and may need training in certain tasks and procedures. If an employee does not use an oxygen meter properly, for example, the data recorded will be useless or, worse still, misleading.

7.3.3 Life-cycle assessment

Life-cycle assessment is similar to environmental impact assessment and to environmental auditing, but it focuses solely on a particular product (Welford and Gouldson, 1993). It is also referred to as cradle-to-grave assessment. It is usually applied to manufactured goods – washing machines, cars, chemical products – but there is no reason why it cannot be usefully applied to aquaculture products. It is a cyclical process designed to lead to continuing improvements in the environmental performance of a product (Figure 7.2). Once the scope of the study has been

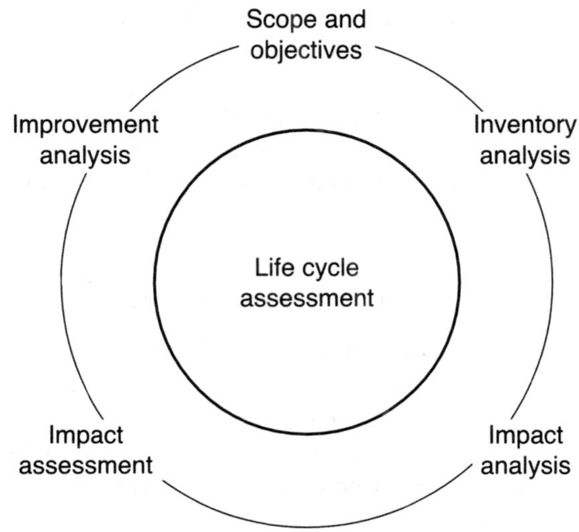

Figure 7.2 Life cycle assessment. (From Welford and Gouldson, 1993.)

defined (e.g. to product X, or products A, B and C), then an **inventory analysis** can be performed. This is a data-gathering exercise and entails the collection of information concerning all the material and energy inputs to a product throughout its life.

For a typical intensive farming operation, for example, one would consider the energy and nutrient content of the fish feed, and also of the wastes in the fish farm wastewater. One would consider the chemical nature of the pharmaceuticals and cleaning agents used in the fish production process. The energy costs and pollution associated with transporting materials to the fish farm and transporting fish from the farm would need to be quantified.

The next stage involves the estimation of the environmental impact of these material inputs and outputs, and energy consumption. This involves two processes: firstly, the **impact analysis**, in which the nature of the impact of a particular material is determined; and secondly, the **impact assessment**, in which each impact is quantified. Having done this (no mean feat), one can now identify those areas of the production process that need most attention, and then develop strategies to improve environmental performance.

7.3.4 Green marketing

Green marketing is really no different from conventional marketing. It is differentiated only by virtue of the message put across by the business to the customer, i.e. 'This business makes every effort to minimize the impact on the environment of its activities, and is committed to continuous improvements in its environmental performance. Our products, therefore, are environmentally friendly.' The obvious similarities with the environmental policy statement in Box 7.1 are not coincidental.

Green marketing still requires thorough market research, appropriate product pricing and promotional strategies, and maintenance of product quality – all the essentials of conventional marketing. One of these is to gain customer confidence in a product, and thus to enhance customer loyalty and repeat sales. This cannot be achieved if the promotional claims for the product do not reflect its real attributes. This is especially so in the case of marketing environmentally friendly goods, where the credibility of the product and, increasingly, of the whole business is of central importance. When it first became fashionable to market products as environmentally friendly, many products carried spurious or misleading claims (Welford and Gouldson, 1993). When these were uncovered by environmental organizations, such as Friends of the Earth with their 'Green Con Award', an attitude of cynicism to green products was created amongst consumers.

To gain credibility for its environmental claims a business must be fully committed to improving its environmental performance in all its activities (Aldridge, 1994). Environmental improvement must be at the heart of company policy and development strategy. An environmental management system, therefore, is essential to the development of a successful 'green' marketing strategy.

7.4 STRATEGIC CONSIDERATIONS

In recent years fish farming has been investigated with regard to the environmental impact of its activities. Most notable concerns included the enrichment of natural waters (both marine and freshwater) and the release of chemicals into the environment (e.g. organophosphates used for sea lice control; use of antibiotics in shrimp farming in the tropics). The aquaculture industry is in a fortunate position, in comparison with many other industries, in that its products and production methods are not inherently damaging to the environment. The damage that does occur can be minimized using current technology and knowledge. Many of the techniques are, in fact, outlined in this book. It is therefore entirely possible for the aquaculture industry to overcome this negative image related to environmental impact, by the promotion and adoption of more environmentally responsible practices.

With environmental policies at its heart the aquaculture industry would have a competitive advantage over many other food industries, many of which would find it much more difficult to change existing production methods to improve the environmental quality of their products. The sea fishing industry, for example, will have to undergo enormous restructuring world-wide if it is to bring catch rates down to sustainable levels and thus become more environmentally friendly.

Runoff from agricultural land (both arable and pasture) is a significant problem, leading to the enrichment of rivers, lakes and coastal waters with nutrients, and to contamination of the environment with pesticides. This runoff is non-point-source in nature, and so cannot easily be dealt with. The typical intensive fish farmer, on the other hand, can install an effluent treatment system and change to a low pollution diet, achieving significant improvements in environmental performance – literally overnight.

SUMMARY

- Environmental management systems offer a strategic framework for the control of environmental impact from aquaculture systems, and for achieving improvements in environmental performance of aquaculture businesses and development programmes.
- The key elements of the environmental management system are:
  - the environmental review;
  - the environmental policy;
  - system design and implementation;
  - the environmental audit.
- The potential benefits of the implementation of environmental management systems include:
  - reduced operational costs;
  - 'green' marketing opportunities.

CHAPTER 8

Future developments in waste minimization

If one considers a production process from an ecological perspective, wastes are simply redistributed resources. In business terms wasted resources indicate inefficiency and lost potential profit. The nutrients and energy in the wastewater from an intensive fish farm might be considered in this way. As environmental concerns gain ground, and as the 'polluter pays' principle becomes an increasingly important policy instrument for environmental protection, the commercial success of fish farms will depend increasingly on their success in either minimizing the production of waste (Chapter 4) or utilizing waste as inputs to other production processes. It is this latter idea that will be developed in this chapter.

8.1 INTEGRATED FARMING SYSTEMS

The integration of aquaculture with other farming activities is not a new idea – it has been going on for centuries in Asia where aquaculture has, in some areas, been comprehensively integrated with other farming systems (Little and Muir, 1987). The integration of crop and livestock farming with fish farming has been successful because of the complementary nature of the input and output characteristics of the different systems (Lightfoot *et al.*, 1993). This is illustrated more fully in Figure 8.1. The output from one part of the system becomes an input to another part of the system and in this way the best use is made of the nutrient and energy inputs to the farming system, so minimizing nutrient leakage (potentially polluting wastes) and potential economic loss.

The key to successful integration is to identify aspects of different farming systems that are complementary and can be integrated. The integration may be achieved with respect to energy, nutrient dynamics and also other resources, such as water. One approach to the development and analysis of integrated farming systems is that of integrated resource management (IRM) (Dalsgaard *et al.*, 1995). IRM uses the ecosystem characteristics of diversity, cycling, stability and (carrying) capacity as the basis for system analysis.

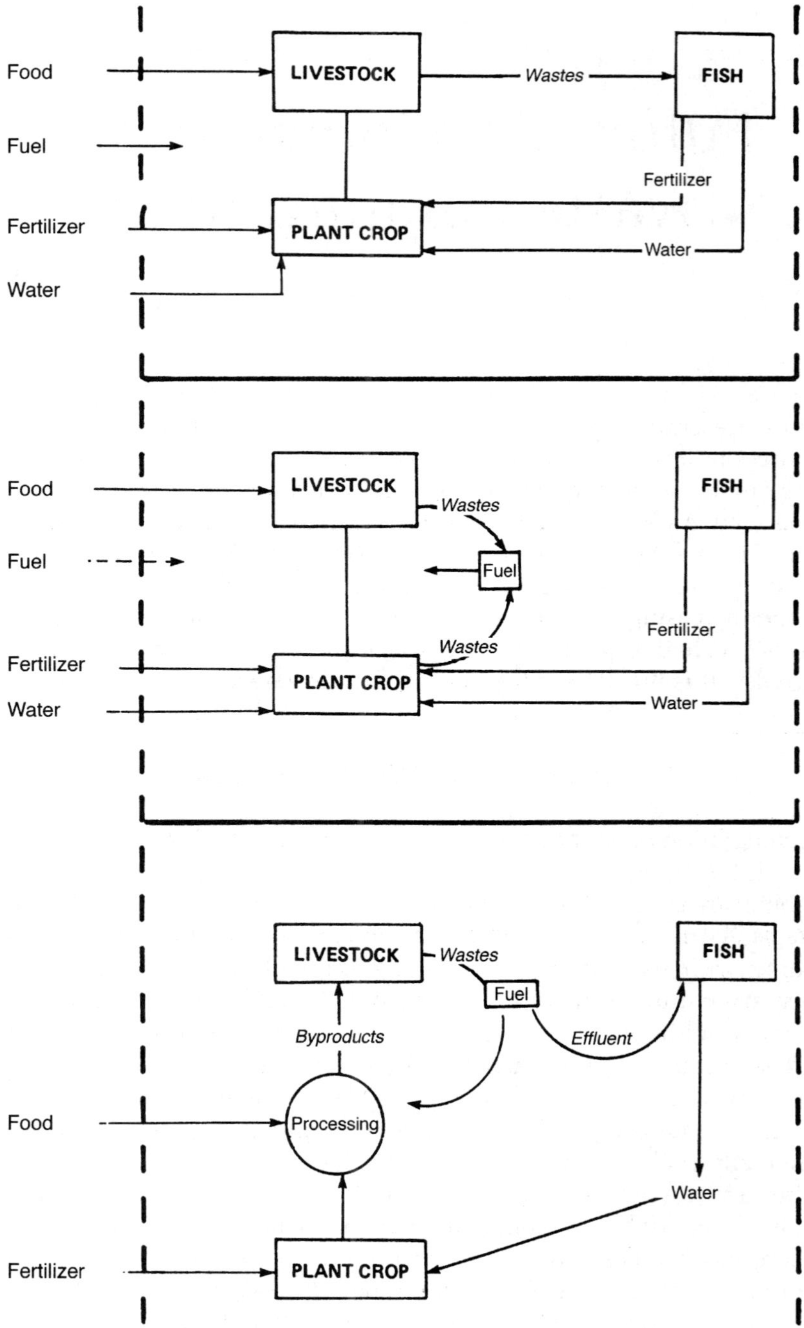

Figure 8.1 Variations on livestock production integrated with fish culture. Note the complementary resource flows, indicated by arrows. (From Little and Muir, 1987.)

- **Diversity**. The maintenance and promotion of diversity is regarded as an important goal in the integration of farming systems. High diversity can help to minimize risk, conserve genetic resources for future exploitation, enhance the efficiency of resource use (e.g. polycultural versus monocultural pond aquaculture) and aid disease control.
- **Cycling**. Recycling of nutrients is a vital process in all ecosystems, contributing to stability and conservation of resources. Many modern monocultural farming systems are characterized by high nutrient leakage (in addition to the unavoidable losses of cropping) and low recycling. They rely on considerable external inputs for their continued viability.
- **Stability**. Some ecosystems are considered to be more resistant to change than others, particularly under conditions of environmental stress. The degree of resilience exhibited by an ecosystem is thought to depend on the nature of the interactions and flows between the species, and the stabilizing or destabilizing elements of feedback loops and interactions.
- **Capacity**. This term refers to the capacity of the environment to support life, and is evident in levels of productivity. Proper management of the farming environment can support and enhance the capacity of the life-support environment.

A computer model, ECOPATH II, has been developed (Christensen and Pauly, 1992) which can be used to characterize farming systems with respect to these four indicators of sustainability. This allows an analysis of the degree of sustainability of a farming system (Dalsgaard *et al.*, 1995), and may indicate the potential for integration between different systems. Figure 8.2 illustrates the results of an evaluation of farming systems in the Philippines (Dalsgaard *et al.*, 1995), the more complex integrated system achieving the highest ecological sustainability score.

The principles underlying integrated aquaculture/agriculture farming systems are equally applicable to the integration of aquaculture with other production processes. In either case the initial considerations should focus on which production processes are complementary to intensive aquaculture, and vice versa. Of particular relevance to this book are the questions:

- What commercial processes yield wastes that could be used as inputs to an intensive aquaculture system?
- To what commercial activity could the waste products from an intensive aquaculture system become an input?

Such an approach will identify new commercial opportunities as well as reducing waste and consequent environmental degradation. The following sections reflect the current direction of integration between

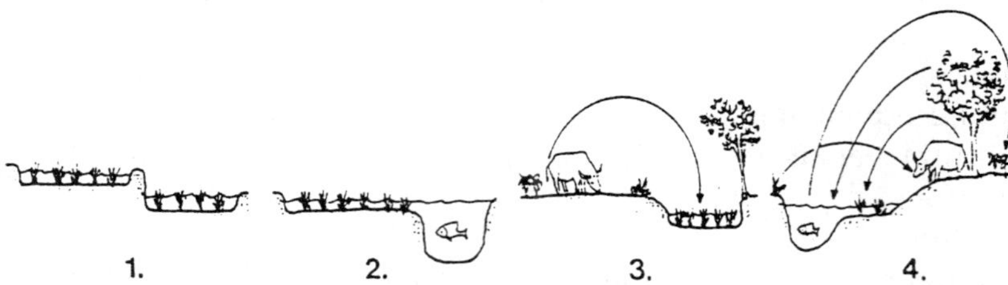

Farming system type	High external input monocropping system		Integrated rice—fish system		Diversified non-integrated system		Low external input integrated resources management system	
Case study enterprise groups	Rice		Rice, fish		Rice, fruit trees, livestock, poultry, vegetables, cereals		Rice, fish, azolla, aquatic plants, livestock, forage grasses, poultry, vegetables, legumes, fruit trees, multipurpose trees	
Ecological sustainability	**Low**		**Medium**		**Medium**		**High**	
Diversity	Low	(0.00)	Low—med.	(0.34)	Med—high	(0.81)	High	(1.26)
Cycling	Low	(—)	Medium	(—)	Low	(0.96)	High	(1.54)
Stability	Low	(0.56)	Medium	(0.65)	Medium	(0.62)	High	(0.68)
Capacity	Low	(0.17)	Low	(0.15)	Medium	(0.49)	High	(0.41)

Figure 8.2 Ecological sustainability across four types of farming systems. (From Dalsgaard *et al.*, 1995.)

aquaculture and other production systems, and point the way for future research and development efforts.

8.2 INTEGRATED MARINE AQUACULTURE SYSTEMS

An integrated farming system based on the natural processes in the marine environment has been proposed as a more sustainable alternative to the monocultural salmon farming industry in temperate seas (Folke and Kautsky, 1989, 1992; Kautsky and Folke, 1990)). This is still based around cage farming of salmon, but also includes other components of the natural food web to mop up the 'leaks' of nutrients and energy from the system.

Figure 8.3 illustrates a chain of production from a coastal ecosystem to the human food economy system. Production from the ecosystem is based on solar energy and nutrients naturally present in the sea. These products (seaweed, mussels and fish in this example) are harvested and processed using fossil fuel energy – the end result being food for human

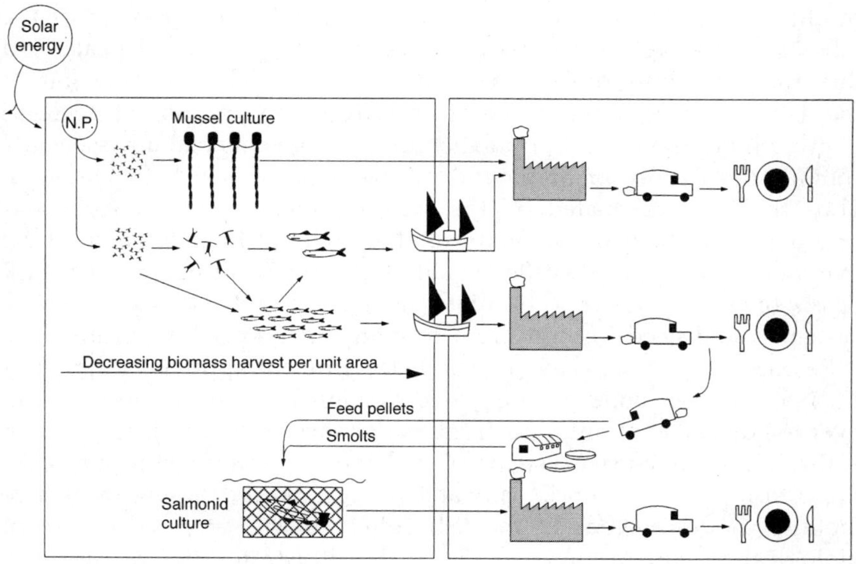

Figure 8.3 The role of mussel and salmonid cultures in the coastal ecosystem and society's economic system. Solar energy and nutrients constitute the ultimate limiting factor for biological production, in this case starting with phytoplankton. Due to energy losses in each transfer between trophic levels, the harvest per area will decrease when a higher level of the food chain is exploited. There are more steps involved in cage farming of salmonids than in culturing of mussels, from solar fixation in the sea until the final product reaches the consumer. (Redrawn from Folke and Kautsky, 1989.)

consumption. In the case of fish farming, some of the production from the sea is processed into fish feeds that are fed to salmon, which are then harvested for human food.

A significant proportion of the salmon feed is lost from the production system through feed wastage and faecal and urinary waste. Ultimately this waste provides nutrients for marine life, but it can cause localized problems because of its concentrated nature (Chapter 2). This problem could be alleviated and turned to economic advantage by siting mussel and seaweed culture alongside salmon cage farms. The seaweed would remove some of the excess nutrients from the water; the mussels would filter out the solid wastes from the fish farm, and benefit from the higher concentrations of phytoplankton in the waters around salmon cage farms. Mussels could be recycled as fish feed to reduce the inputs of fishmeal, which has a high embodied energy as a considerable amount of fossil fuel energy is used in its production.

Trials carried out in the Baltic have demonstrated the potential of an integrated open water system (Welsch, 1992). In this study, the fouling of cage nets by blue mussels was allowed, and the effects of this on the production of rainbow trout were monitored. The main effect of the mussels on the nets was to reduce water exchange within the cage and thus the availability of oxygen. At times, therefore, the feeding rate of the fish had to be reduced. One would expect a concomitant decrease in growth but the mussel population on the cage nets had an associated community, mainly of amphipod crustaceans, on which the fish fed. Thus the FCR, calculated on the consumption of pelleted feed, was improved and the fish benefited from the high nutritional quality of the live food, having excellent flesh quality and colour at harvest. The risk of dissolved oxygen concentrations falling to critical levels in such a system would need to be addressed in any commercial application.

Researchers in Israel have developed a different approach (Shpigel *et al.*, 1993). Closed pond or tank-based systems offer much more control over the dispersal of nutrients from fish or shrimp culture facilities than is the case for open sea systems. They have integrated, on a pilot scale, the culture of sea bream in saltwater ponds, and the culture of bivalve molluscs and seaweeds in the fish culture wastewater, as a form of biological treatment process. Figure 8.4 illustrates the relationship between the different components of the system.

The fish ponds were managed intensively. Pelleted feeds provided the nutritional requirements of the fish. The water was aerated and exchanged at a rate of 50% per day. From the fish tanks the water drained into a sedimentation pond, which had a residence time of approximately 1 day. The sedimentation pond was used to supply water and nutrients to a bivalve culture unit: water was continuously pumped to a tank containing oysters and clams arranged on suspended trays at a density of 25–50 kg m^{-3}, and discharged back to the sedimentation pond. The effluent stream continued from the sedimentation pond to another bivalve culture unit, the primary purpose of which was solids removal. It then continued to a seaweed culture unit containing *Ulva lactuca*, the water lettuce, and thence to waste.

The only nutrient input to the system was the pelleted fish feed. Figure 8.5 shows the path of the nitrogen inputs through the system. Note that the nitrogen content in the final effluent is only 4% of the input loading, compared with a conventional marine cage farming system in which about 70% of the input nitrogen is discharged to the environment.

Shpigel and Neori (1996), in a variation of the system described by Shpigel *et al.* (1993), proposed a modular system for integrated mariculture using seaweed and abalone as the basic components and increasing the complexity by adding fish and clam culture systems. The predicted performance of the various permutations of the system were based on data collected in studies of the individual components as isolated systems.

In the core module, seaweed would be cultured in fertilized saltwater ponds, and the harvested plant biomass would be fed to abalone. The

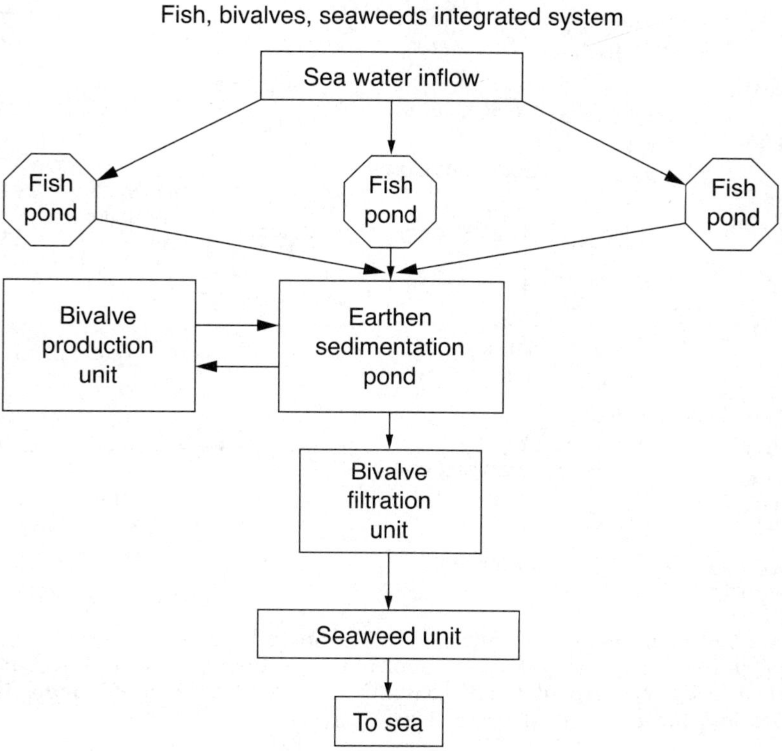

Figure 8.4 Proposed integrated marine aquaculture system. The combination of different production processes maximizes the utilization of system inputs. (From Shpigel *et al.*, 1993.)

proposed second stage of integration would be the introduction of fish culture to the system. Wastewater from fishponds would be treated in a lagoon planted with seaweed; most of the treated water would be recycled to the fish pond. The abalone module would be connected to the seaweed treatment lagoon in the same way. Seaweed biomas would be harvested as food for the abalone. For stage three of the integration the addition of a bivalve culture module is proposed. Here fish feed would remain the only source of nutrients for the whole system. Fish culture wastewater would flow into a sedimentation pond, where micro-algae would be responsible for most of the water purification. Clams (*Tapes philippinarum*) would be cultured in this module, and would utilize the micro-algae as food. Downstream of the clam module would be two water-polishing modules, the first containing more clams and the second containing seaweeds (e.g. *Ulva lactuca, Gracilaria* spp.). The seaweed biomas would be used to support the abalone module. However, the reduced supply of nutrients to the seaweed module would limit its

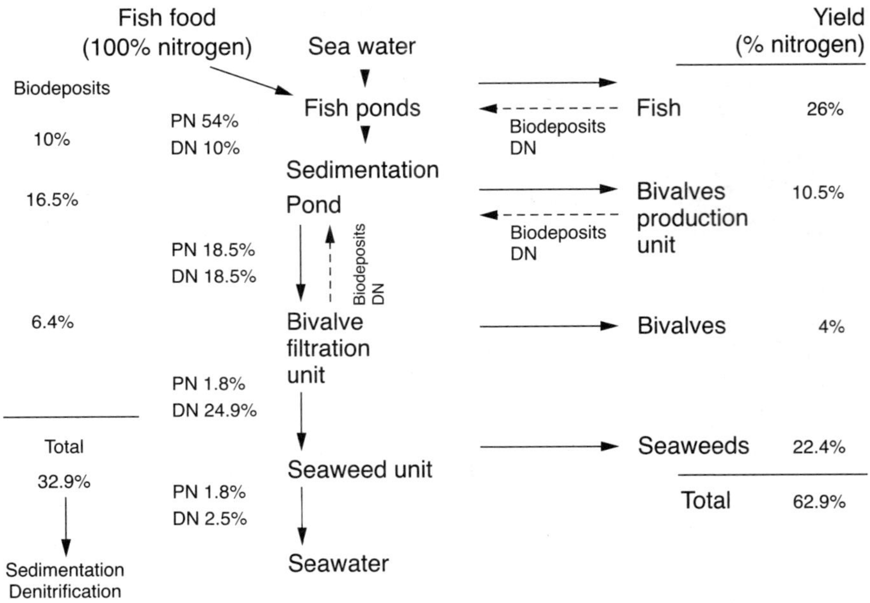

Figure 8.5 Flow diagram of nitrogen in a proposed integrated marine aquaculture system. All values are percentages of the nitrogen introduced in the fish feed. **PN**, particulate nitrogen; **DN**, dissolved nitrogen; **biodeposits** = faeces and uneaten food. (From Shpigel *et al.*, 1993.)

productivity and the abalone unit would therefore only supply a secondary crop.

Recently the culture of the red alga *Gracillaria* spp. has been investigated in an effort to find a solution to the serious pollution problems posed by intensive shrimp farms in the tropics (Funge-Smith and Briggs, 1994). *Gracillaria* is an important crop, being popular locally as a food, and in considerable demand internationally as a source of agar. This substance is used primarily for microbiological work in hospitals and laboratories.

Figure 8.6 illustrates a typical nutrient budget from a tropical intensive shrimp farm. On an individual basis it compares favourably with that of a salmon or trout cage farm. The environmental problems associated with shrimp farming arise from the high density of farms in coastal regions; the concentration of nutrients in the coastal waters of shrimp farming areas is consequently very high.

The trials described here were conducted in Thailand, on an intensive shrimp farm (Funge-Smith and Briggs, 1994). The shrimp were grown in static water ponds, where the growth of phytoplankton was encouraged to maintain the water quality (by removal of ammonia from the water). Periodic water changes were necessary, especially as the biomass

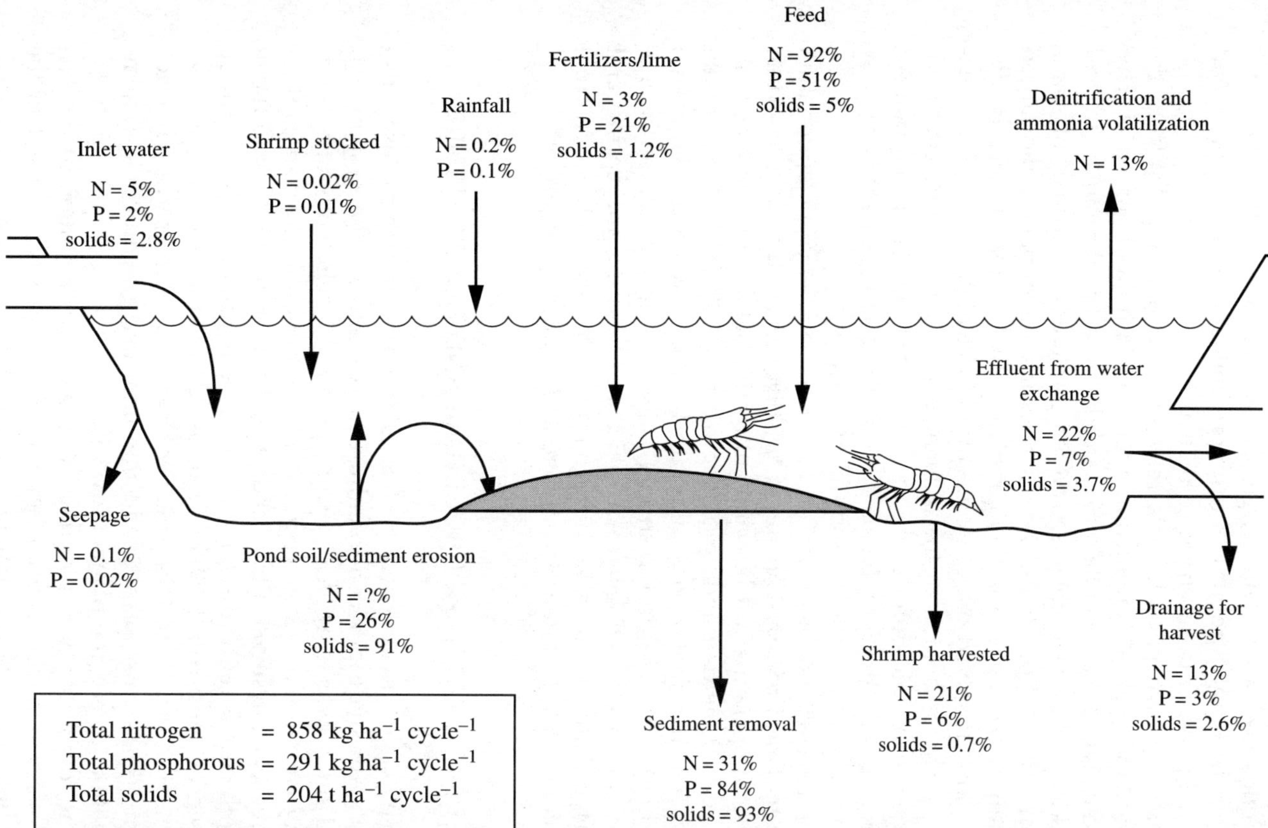

Figure 8.6 The sources and sinks of nitrogen (N), phosphorus (P) and solids in intensive shrimp ponds. The numbers refer to the percentage of the total either introduced to or exported from the pond, as indicated by the direction of the arrow. (From Briggs and Funge-Smith, 1994.)

of shrimp in the pond increased and the pond was completely drained at harvest. Some farms have settlement ponds, to remove suspended solids from the influent as well as the effluent water. To be effective these ponds need to be quite large (Chapter 5). A settlement pond for the water discharged from the farm represents unproductive space, and this is a significant disincentive to their use.

Gracillaria grown in tank trials removed ammonia rapidly from the water, achieving an 82% reduction after only 18 hours in one trial. This was indicative of the potential of seaweeds for nutrient removal. In field trials, some practical problems were encountered. Settlement of solids on to the seaweeds was particularly serious in effluent ponds, and caused the death of the algae. Another problem, again in the effluent ponds, was the low dissolved oxygen concentration at night: the high concentration of phytoplankton in the effluent ponds deprived the *Gracillaria* of oxygen for respiration during the hours of darkness. In the influent settlement ponds, these problems did not occur and the *Gracillaria* grew well and removed nutrients from the water.

Other research has also demonstrated the efficiency of ammonium uptake by *Gracillaria*. Buschmann *et al.* (1996) found that in a recirculation system operating in Chile and cultivating salmonids (*Oncorhynchus kisutch* and *O. mykiss*) and *Gracillaria chilensis*, 50% of the ammonium produced in the system was taken up by the seaweed in winter and approximately 90–95% in the spring. This was under conditions where ammonium concentrations reached 500 $\mu g\, l^{-1}$ during the spring and summer months. Removal of nitrogen, phosphorous and carbon from the system through different components is shown in Figure 8.7.

8.3 INTEGRATED AQUACULTURE AND HORTICULTURE

Jungersen (1991) investigated the feasibility of integrating intensive fish culture with hydroponic horticulture in Denmark. The system was centred on a heated, recycle system producing eels. Although solids removal and ammonia conversion to nitrate are established in conventional recycling systems, nitrate and phosphate removal is more difficult and both are usually discharged to the environment. The solution proposed by Jungersen (1991) was to use this nitrate and phosphate as nutrients for plant growth. The fish farm wastewater was diverted through a glasshouse plant production unit, using hydroponic systems to irrigate and feed the plants (Box 8.1).

It was found, for example, that the nitrate concentration in the fish farm wastewater was reduced by about 60% when it was used in the hydroponic production of tomatoes (Jungersen, 1991). Large-scale trials suggested that the production of 100 t of eels could support a production of 870 t of tomatoes. The flow of nutrients is illustrated in Figure 8.8. It

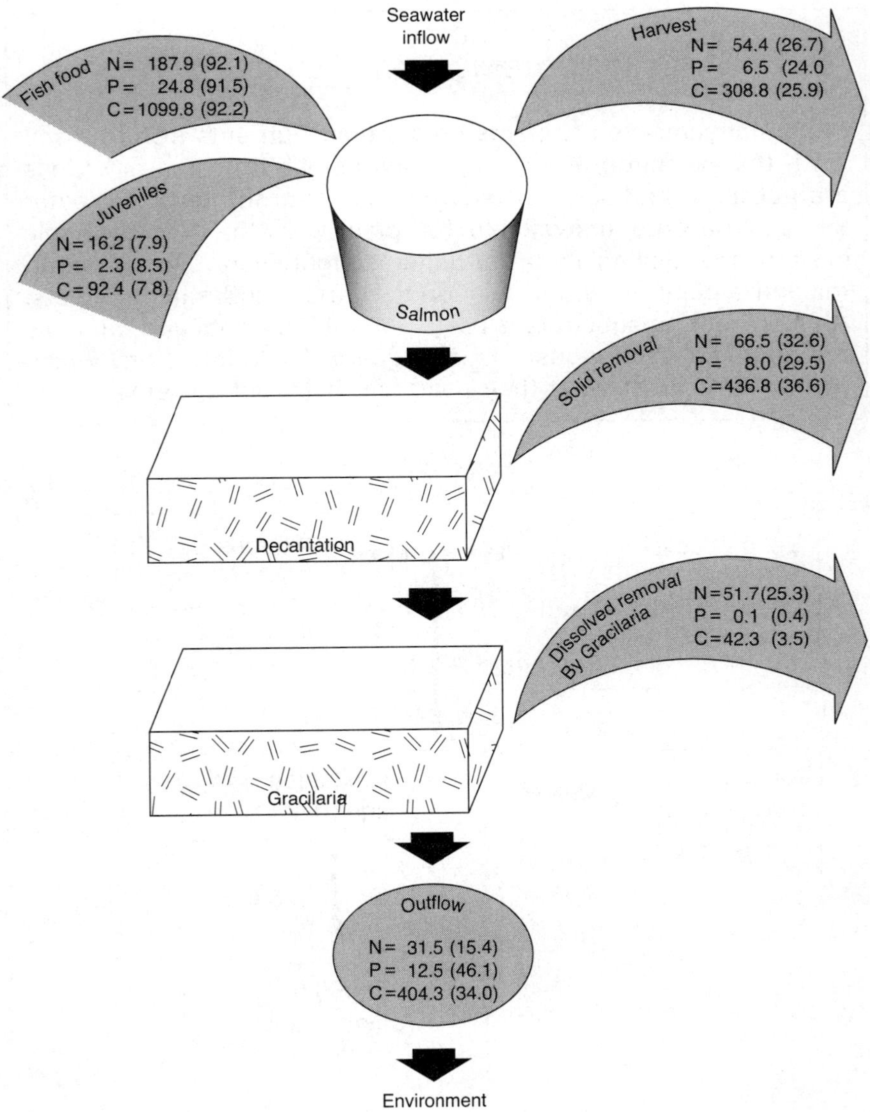

Figure 8.7 General nutrient budget, kg (%), during total production period (13 months) obtained for integrated 1.6 t intensive salmon/*Gracillaria* farm. (From Buschmann *et al.*, 1996.)

should be noted that the carbon dioxide requirement of the plants is also partially met by the discharge of carbon dioxide from the fish farm unit.

McMurty *et al.* (1993a,b) demonstrated the potential for integration of a fish culture unit with a hydroponic cultivation of tomatoes. In this

BOX 8.1
HYDROPONIC HORTICULTURE

In natural ecosystems, plants obtain their nutrients from the soil (with the exception of carbon, which comes from the air). Soils are not inert, and some of the minerals and nutrients that enter the soil become unavailable for plant growth. In hydroponic systems the availability of nutrients is highly controlled, so as to maximize plant growth in relation to nutrient and mineral inputs. An inert rooting substrate is used, and the nutrients and minerals are supplied continuously in an aqueous solution, from which plants can extract what they need via their root systems.

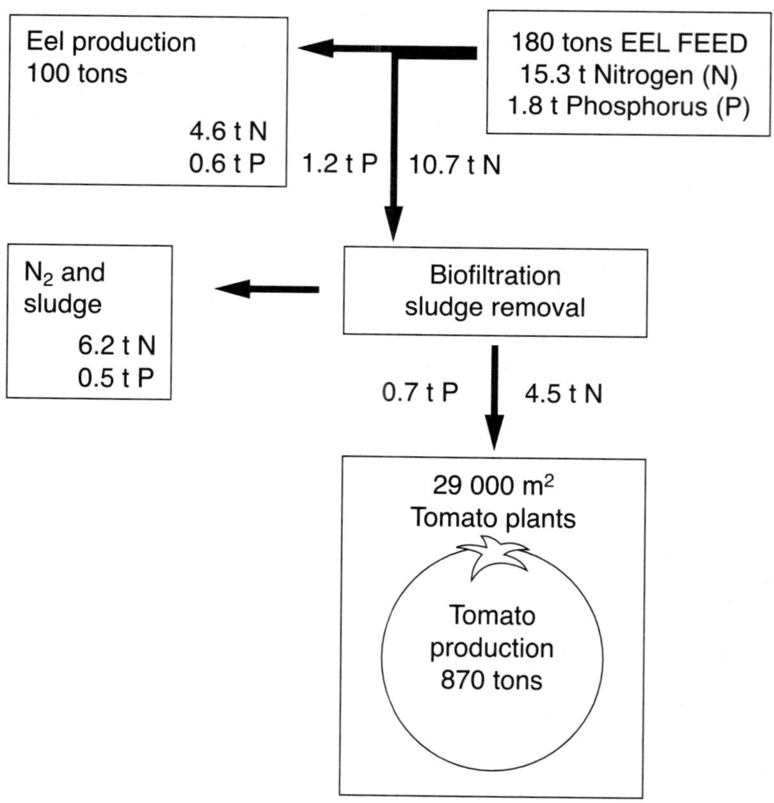

Figure 8.8 Flow of nitrogen and phosphorus in an integrated eel/tomato production system. (From Jungersen, 1991.)

example effluent water from a hybrid tilapia (*Oreochromis mossambicus* × *O. niloticus*) unit was recirculated through sand biofilters in which were planted tomato plants. Both systems were managed successfully and yielded acceptable crops. It was found that each 1 kg increase in fish biomass provided sufficient nutrients from metabolic processes to support the growth of two tomato plants for 3 months.

8.4 TOWARDS ZERO EMISSIONS

With the goal of sustainable development in mind the United Nations University conceived the Zero Emissions Research Initiative (ZERI) (Pauli, 1994; Wenhua, 1995). The basic concept of ZERI is that industry will have to re-engineer manufacturing towards the objective of zero emissions, in order to achieve sustainable development. The process of elimination of all forms of waste will bring together sectors that seem to have little in common into clustered production systems. In these systems, today's waste products become inputs to other, adjacent production systems.

As part of the ZERI initiative, certain industries were identified around which a research programme could be developed. One such was the brewing industry, which has a significant waste problem. Spent grains and wastewaters from the brewing process are rich in organic nutrients. Traditionally spent grains are disposed of in landfill sites or used in animal feeds. Wastewaters are discharged to the natural environment.

The challenge for ZERI was to identify other processes that could be integrated with the brewing process to achieve zero emissions. A number of pilot projects have been established around the world to develop regionally appropriate solutions to this challenge. The European pilot project is planned to be sited at the Earth Centre in the UK. This is a new Sustainable Futures centre, devoted to education and research. Here a micro-brewery has been chosen as the core unit (S. Bedford-Clark, personal communication). Downstream of this, utilizing the brewing wastes, will be fish culture, algal culture, hydroponic horticulture and biogas digestion. This will be housed in an energy-efficient building to enable the maintenance of higher than ambient temperature. Thus the efficiency of the production processes will be optimized.

The project is still in the planning stages at the time of writing, but a technical feasibility study using the systems ecology approach has confirmed the suitability of these production processes for integration. The criteria used were: volume and concentration of nutrients and energy; compatibility between the scale of each production system (based on the amount of inputs available); and compatibility of each system to the overall process flow (bearing in mind seasonal and environmental demands). A flow diagram of the system is presented in Figure 8.9.

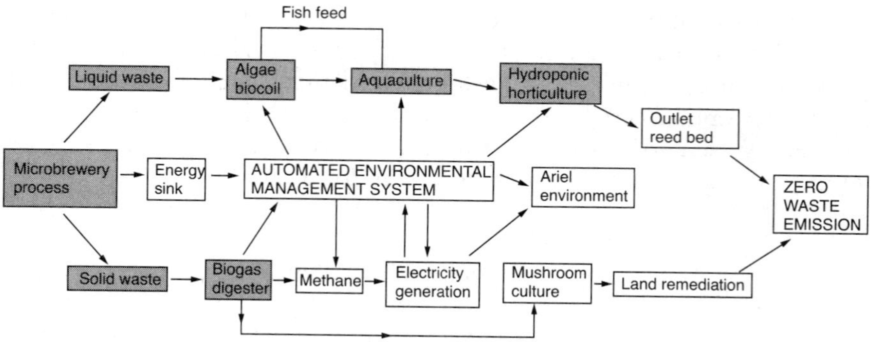

Figure 8.9 Conceptual flow diagram for an integrated production system designed to minimize waste. Shaded boxes indicate components operating within an energy efficient building. (Personal communication from Bedford-Clark.)

8.5 DEVELOPMENTAL CONSTRAINTS ON INTEGRATED PRODUCTION SYSTEMS

There are some important constraints to the development of integrated systems such as those discussed in the preceding section, namely:

- system optimization;
- chemical usage;
- production intensity;
- synchronicity;
- proximity.

8.5.1 System optimization

In monocultural farming systems, maximization of productivity is the normal goal (assuming that productivity equates to profits). In integrated production systems, the maximization of productivity in each component is not necessarily the best way to achieve the optimum performance of the system overall. One needs instead to optimize (as opposed to maximize) productivity of each individual component **bearing in mind the constraints imposed on each by the needs of the other components of the system**. In this way the optimal performance for the system as a whole will be achieved. An integrated production system is, essentially, a whole new system to be operated as one, rather than a collection of separate operations.

8.5.2 Chemical usage

The use of chemicals in the fish culture cycle is a particular problem, especially when downstream processes produce food products for human consumption. Organophosphates, for example, used in the control of sea lice, are not compatible with mussel culture. This is both from a biological viewpoint and with regard to human health. A different approach to disease control may be needed in integrated systems, in which the optimization of the system as a whole is the primary aim. This might lead, for example, to the acceptance of a higher rate of fish mortality than normal, so as to reduce chemical use and the effects of that on the other components of the system.

8.5.3 Production intensity

The compatibility of different production processes must be evaluated with regard to the rate of flow of nutrients through the system, as well as to nutrient dynamics. In the fish/horticulture example the area required for tomato production is about 10 times larger than that for the fish culture. A typical commercial grower in Europe might have 1–2 ha of glasshouse and the opportunities to combine this with commercial fish culture would be limited. However, a large horticulture production unit might comprise up to 8 ha of glasshouse, making integration with intensive fish culture a more feasible proposition.

8.5.4 Synchronicity

The compatibility of the life cycles of organisms in biological productions systems is important. Generally this is likely to be more of a problem in higher latitudes, where some species may be more affected than others by seasonal factors (such as temperature and light intensity). In addition to biological constraints, there may be seasonal market constraints, which may affect the viability of an integrated production system.

8.5.5 Proximity of the component processes

In open sea systems, proximity of the components of the integrated system will be important. The further the downstream components are from the primary source of nutrients, then the greater is the likelihood that nutrients will have been dispersed by tides and currents and are no longer available to the integrated system.

8.6 CONCLUSION

The integration of production processes to achieve improvements in environmental performance is an emerging theme. Aquaculture is inherently suited to integration with other production systems based on environmental goods and biological processes.

The technological constraints are developmental rather than fundamental. The main constraints are economic factors, and outdated attitudes to industrial development. The adoption of sustainability as a key developmental goal will, perhaps, remove these constraints.

SUMMARY

- Integrated farming systems which include aquaculture are common in certain regions and have a long history.
- Integrated farming systems aim to optimize the efficiency of resource use across a whole range of interconnected farming activities. This is an effective waste minimization strategy.
- Integrated production systems are advocated as a means of achieving greater sustainability of aquaculture.
- Integrated production systems in aquaculture include:
  - marine fish, seaweed, mollusc culture;
  - shrimp and seaweed culture;
  - fish and horticulture;
  - aquaculture and brewing processes.
- Development constraints include:
  - system optimization;
  - chemical usage;
  - production intensity;
  - synchronicity of processes;
  - proximity of processes.

CHAPTER 9

Environmental protection and aquaculture

The regulation of environmental quality has traditionally taken a single-medium approach; the controls of air, water or soil pollution have been conducted under separate legislation and, more often than not, by separate controlling authorities. In the last decade this approach has increasingly been seen as inappropriate. A widely held view is that it has resulted, in many cases, in the pollution problem being diverted from one medium to another, rather than being effectively resolved. A more integrated approach, aptly referred to as integrated pollution control, stresses the need to find solutions to pollution problems that result in the least pollution measured across all three media.

Associated with this change in attitude by pollution control authorities has been the change in emphasis from 'end-of-pipe' technologies (i.e. filtration equipment, etc.) for environmental protection, to the development of waste minimization strategies. A strong force behind this movement is the 'polluter pays' principle, the implementation of which is making industrial waste disposal increasingly expensive. Reducing the cost of disposal can be achieved either by installing effluent cleaning technology (which tends to be costly) or by adopting waste minimization strategies. Taking into account the efficiency and material input savings often made through implementing waste minimization programmes, the economic incentive is clearly in their favour.

Regulatory authorities and the aquaculture industry are taking a more integrated approach to pollution control and environmental protection but, against this background, much environmental protection legislation still conforms to the traditional single-medium approach. Over the next few years this will inevitably change, and more integrated approaches to the regulation of the environmental impact of aquaculture will be adopted around the world.

This chapter presents a brief review of the environmental protection legislation applicable to the aquaculture industry, followed by an appraisal of significant recent developments. Finally, a strategic approach to environmental management for this sector is proposed.

Table 9.1 Summary of European environmental regulation of aquaculture (modified from FES, 1992)

Country	*Limits to production*	*Mandatory water treatment*	*N and P loading limits*	*Low pollution diets specified*	*FCR standards specified*	*Notes*
Austria	N	Y	Y	N	N	
Belgium	N	N	N	N	N	Mostly extensive production methods
Denmark	Y	Y	Y	Y	Y	
Finland	Y	Y	N	N	Y	
Germany	N	N	Y	N	N	Mostly extensive production methods
Iceland	Y	Y	N	N	N	
Ireland	Y	Y	Y	N	N	
Italy	N	N	Y	N	N	
Norway	Y	–	N	N	Y	
Sweden	Y	–	N	N	N	
United Kingdom:						
England and Wales	N	Y	Y	N	N	
Scotland	Y*	N	Y	N	N	*Cage farms only
Shetland	N	–	Y	N	N	
Ireland	N	N	Y	N	N	

–, no information.

9.1 EXISTING APPROACHES TO THE REGULATION OF POLLUTION FROM AQUACULTURE

The following case studies have been chosen to illustrate the range of environmental protection legislation affecting the aquaculture industry.

CASE 9.1: EUROPEAN UNION AND SCANDINAVIA (FES, 1992; European Commission, 1995)

A survey of the environmental protection regulations governing the aquaculture industry within the EU illustrates most of the common features of environmental protection in aquaculture throughout the world. The range of regulations and their application is summarized in Table 9.1.

A policy instrument common to many countries is the setting of limits on nutrient loading, and a statutory requirement for water treatment prior to discharge. However, the implementation of these two regulations differs widely between states. In the UK the requirement for wastewater treatment is usually limited to settlement ponds, with no stringent guidelines on their design; in other countries, physical filtration equipment may be specified in addition to settlement ponds.

With regard to limits on nutrient loading, these too differ widely in their implementation. Table 9.2 illustrates typical differences between Denmark and England in this respect.

Production limits are also relatively common but are usually applied only to floating cage farms, as an alternative to limits on nutrient loading (which cannot be effectively monitored in these cases). Denmark is a notable exception to this observation: there, production limits are set for land-based trout farms. This measure is designed to prevent increases in production as improvements in feeding management reduce pollution loads in the wastewaters. Thus one of the main short-comings of consents to discharge pollution is overcome.

Table 9.2 Typical discharge consent limits on nutrient loading from land-based salmonid farms – values (mg l^{-1}) indicate maximum allowable increase of outflow concentration over that of inflow (from FES, 1992, and National Rivers Authority, UK)

Nutrient	*Denmark*	*England*
BOD	1.0	2.0
Suspended solids	3.0	5.0
Phosphorus	0.05	–
Ammonia nitrogen	0.4	0.5
Total nitrogen	0.6	–

BOX 9.1
A CRITIQUE OF EIA

A major flaw in the EIA (environmental impact assessment) system is that it reacts to specific development proposals – it does not anticipate them. This is particularly relevant for some sectors of the aquaculture industry: shrimp farming, for example (Chapter 2), where rapid sector development at a regional level has caused significant environmental problems for the farmers themselves. The significance of this distinction is that EIA cannot steer developments towards locations more resilient to the environmental impacts of the project. Nor does it consider the cumulative impacts of more than one project. EIA is project specific. As most EIAs are undertaken by the developer, cumulative impacts of other developments are rarely taken into account (Therivel *et al.*, 1992). Cumulative effects are especially important in the aquaculture industry. It is common to have more than one development using the same resource – a river, lake or coastal zone. The cumulative effects of such developments can have serious consequences, not just for the wider environment but also for the fish farmers (Chapter 2).

The alternative options for development are usually addressed in only a limited way in EIA. Often a project has been planned in considerable detail, and irreversible decisions taken, before an EIA is undertaken (Therivel *et al.*, 1992).

In addition to the regulations concerning the operational management of fish farms, there are EU regulations concerning environmental impact assessment (EIA). These require that an environmental impact statement (EIS) be prepared for certain developments and submitted to the planning control authorities. The EIS should detail the potential environmental impacts from the development and propose appropriate mitigation measures. It should also examine alternative ways of achieving the goals of the development and explain why these have not been adopted. In the EIA Directive, an Annex lists two classes of development: those for which EIA should be compulsory, and those for which the requirement for EIA is at the discretion of individual member states. Aquaculture falls into the latter category, and is usually only required for larger developments (e.g. salmonid farms of 100 t per annum production in the UK). Most aquaculture developments fall outside this measure (Howarth, 1990). The implications of this failing are summarized in Box 9.1.

To summarize, the range of policy instruments in place in the EU includes:

- development permits, specifying size and/or location of the fish farm;
- environmental quality standards (achieved through nutrient loading limits and other conditions specified in consents to discharge);
- waste minimization regulations, relating to feeding management;
- discretionary requirement for EIA.

With very few exceptions, these measures have evolved from general environmental legislation in an *ad hoc* way. They come nowhere near an integrated approach for environmental management of the aquaculture industry. The two notable exceptions to this are Norway and Denmark, which can claim to be leading the way in environmentally sound management of aquaculture development. Their approaches to environmental protection in aquaculture are summarized in Cases 9.3 and 9.4.

CASE 9.2: BRITISH COLUMBIA, CANADA
(Caine and Castledine, 1991)

British Columbia, like Scotland and Norway, has seen a rapid development of marine salmon farming in floating net cages in the last decade. The need to control development to prevent environmental damage has been recognized and is reflected in the environmental regulation of the industry.

For new salmon farm proposals a development plan has to be submitted to the Ministry of Agriculture and Fisheries that clearly details the technical aspects of the proposal. This is evaluated with regard to the biophysical characteristics of the proposed site and the potential impact of the proposed development. An important tool in this process is a computer model which is used to produce a site-specific assessment of the potential sediment and nutrient loading on the environment resulting from the proposed level of production, in order to gauge more accurately the potential impact on the environment.

The details of the development plan submitted as part of the application procedure constitute the specific terms of the licence conditions. Monitoring of the conditions involves a biannual aerial survey, plus annual inspection of farm management and development (where these aspects are relevant to environmental protection).

The Ministry of the Environment is responsible for permitting and monitoring waste discharges from fish farms. The regulations specifically requiring an aquaculture waste control programme were introduced in 1988, and are subject to periodic review to evaluate their efficacy. With regard to marine salmon farms, monitoring is undertaken at one of three levels, depending on the annual feed input to the farm:

- Fishfarms using less than 120 t feed per annum are exempt from a waste permit.

- Fishfarms using 120–630 t feed per annum have to submit oceanographic data, an annual dive survey of the seabed under the cages and operational records. They, too, are exempt from a waste permit, provided they comply with the other conditions.
- Large operations, using over 630 t feed per annum, must, in addition to the above, undertake more intensive annual monitoring, including evaluation of nutrients in order to comply with the necessary waste permit.

The development plan for each fish farm is subjected to a limited environmental impact assessment and is used as a baseline reference for operational management standards for the farm. The spectre of the conditions of a waste permit is an incentive for smaller operators to comply with sensible environmental management practice.

CASE 9.3: DENMARK (Jensen, 1991)

The regulatory authorities in Denmark recognized that urgent regulation of the aquaculture industry was needed to bring about rapid reductions in its impact on the freshwater environment. Fish farms were clustered on suitable water courses and cumulative environmental effects were sometimes severe. The traditional methods of environmental regulation were not effective. Water quality emission standards were often seen as a right to pollute. The considerable advances in feed manufacture in the 1980s greatly reduced the pollutant loading from the farming process and fish farmers recognized that they could now increase production without contravening their discharge consents. Trout production in Denmark increased from 12 500 t in 1974 to 24 000 t in 1983, 29 000 t in 1987 and 34 000 t in 1989.

By 1985 Denmark had established environmental quality objectives for all freshwaters. In many cases these were not being achieved, and overdevelopment of aquaculture was seen as one of the main causes for this failure. The need for a new approach was recognized in 1987, and the rationale and methodology for defining it are discussed in the following paragraphs.

The development of low pollution diets, and their obvious environmental benefits, made a **waste minimization strategy** a very attractive environmental policy approach. There are also real economic benefits to the aquaculture industry, including:

- improved feeding efficiency and reduced feed costs;
- lower wastewater treatment costs;
- environmental benefits (reduced risk of 'self-pollution').

A **risk analysis** was undertaken to determine what levels of fish farm productivity posed environmental risks. Analysis of historical river quality data and fish farm records indicated that the relationship

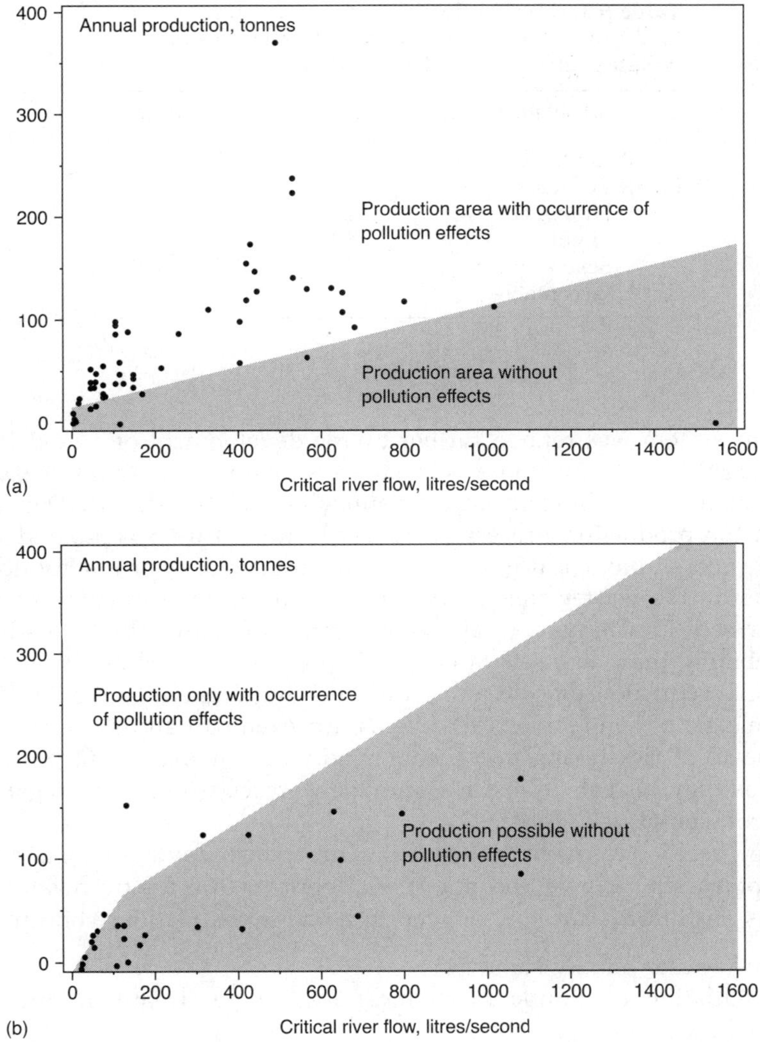

Figure 9.1 Risk analysis for trout farms in Denmark. Those situations falling into the shaded area represent low risk, and were used to define environmental carrying capacity. (From Jensen, 1991.)

between river flow and annual fish production could be used as an indicator of the risk of pollution events. Classification of the level of risk allowed this data to be used as a crude indicator of the environmental carrying capacity of the receiving waters. Figure 9.1 illustrates this process. In this way three classes of risk were defined:

- Class I: Acceptable – production without risk of pollution effects.
- Class II: Risk area – production carries risk of pollution effects.
- Class III: Unacceptable – production will cause pollution effects.

Table 9.3 Distribution of freshwater fish farms and of total trout production in Denmark in classes of pollution effects (from Jensen, 1991)

Classes of pollution effects	*% Farms*	*% Production*
I: Acceptable	36	25
II: Risk area		
No effect	14	14
Slight effect	19	21
Severe effect	10	12
III: Unacceptable	12	28

This system was refined further by the development of a model that was used to assess in more detail those situations falling into Class II. The model uses FCR (calculated on annual data) to indicate BOD loading from the production process at each fish farm. This is compared with the results of biological monitoring programmes of river quality downstream of the wastewater discharge point (using the saprobien system, described in Chapter 6, as an indicator of actual BOD loading). Combining these two sets of data, it is possible to estimate the amount of BOD reduction that occurs within the environment of the fish farm (by means of solids removal systems, for example) and so re-evaluate the level of risk taking into account individual management practices. In this way all fish farms in Denmark were classified with regard to environmental risk (Table 9.3).

New regulations were drafted using this analysis (and others) to define acceptable operational and managerial performance for freshwater fish farms. Fish farms are now obliged to abide by regulations concerning:

- wastewater quality standards;
- installation of simple mechanical wastewater treatment prior to discharge;
- installation of settling ponds;
- regulations concerning feed composition and FCR;
- maximum allowable feed consumption (thereby setting production limits in relation to environmental carrying capacity);
- abolition of stretches of 'dead river' (i.e. over-abstraction).

At a higher level the new regulations provided for:

- regional powers to implement further regulations if local environmental conditions indicate that these are necessary;
- the conditions regarding feed quality to be phased in over 3 years;
- the regulated output (through feed amount limitation) should take account of historical output;

- fish farming representatives and feed manufacturers sit with the regulatory authorities on an advisory board with regard to the implementation of the new regulations.

CASE 9.4: NORWAY – THE LENKA PROJECT
(Bergheim et al., 1991; Ibrekk et al., 1993)

LENKA is a coastal zone management programme specifically designed for aquaculture development. The objectives of LENKA are:

- to ensure continued growth in the aquaculture industry and sensible exploitation of the coast in the future;
- to provide responsible authorities with an efficient tool to localize the growth in fish farming to the most suitable areas within the limits of the environment's natural holding capacity and without causing conflicts with other user's interests.

The coastal zone was divided into ecologically significant zones, based on their physicochemical and biological characteristics. Each zone was then evaluated with regard to its assimilative capacity for the nutrient wastes (principally N) from aquaculture. In such a way a carrying capacity for each zone for aquaculture was determined. Figure 9.2 illustrates the division of part of the Norwegian coast into zones of suitability for cage aquaculture.

The LENKA project has been used to inform the planning process at different levels of government. It has been used to assist in the evaluation of individual aquaculture developments by local authorities. It has been also used strategically in the formulation and implementation of government policy (at national and regional levels) – for example, by assisting in the identification of infrastructure needs for the further development of the aquaculture industry.

9.1.1 Summary of cases

The key feature of the Danish and Norwegian approaches which point the way to the future are:

- waste minimization strategies;
- development controls based on assessments of environmental carrying capacity.

Why are these such important features of their policies? It is because they embody sustainability as a central goal of development, and they recognize the link between development and environmental health. This represents two major steps in the development of a strategic approach to aquaculture development and environmental management, which

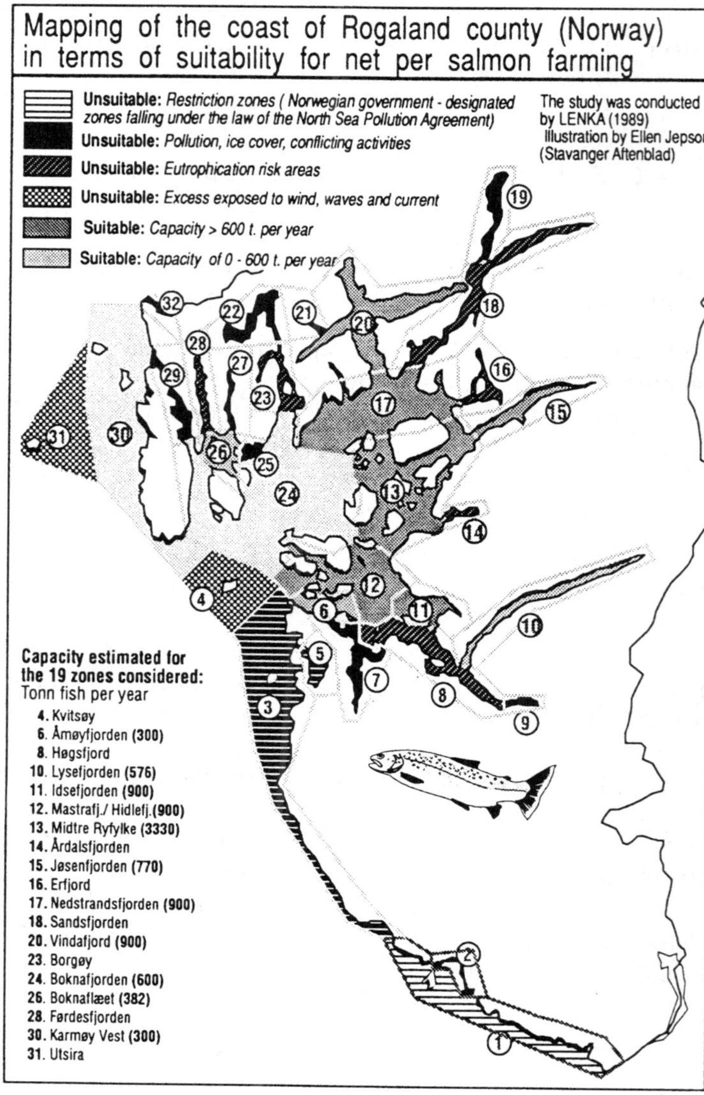

Figure 9.2 The suitability for aquaculture of 19 coastal zones within the county of Rogaland, Norway. (From Bergheim *et al.*, 1991.)

sees individual developments in a wider context. A general methodology for this type of approach has been developed and is referred to as strategic environmental assessment.

9.2 STRATEGIC ENVIRONMENTAL ASSESSMENT (SEA)

SEA can be viewed as environmental impact assessment of institutional policies, plans and programmes (World Bank, 1991; Therivel *et al.*, 1992). It can be applied at local, regional or national levels, but not for individual projects. It is also a means of implementing sustainability. EIA can be of only limited use as an environmental protection tool if the strategic decisions that ultimately generate the projects are intrinsically harmful to the environment, or are implemented without regard to their potential impact. A proactive approach using SEA would foster policies, plans and programmes that were inherently sustainable with respect to the exploitation of natural resources.

Ideally, SEA should be based on sustainability objectives, not development-led objectives. Sustainability in this context would be defined by environmental carrying capacities, from which would emerge threshold limits for development that would not be exceeded (Therivel *et al.*, 1992). Within these constraints other factors (social, economic, institutional) can be optimized.

9.3 A STRATEGIC FRAMEWORK FOR ENVIRONMENTAL MANAGEMENT IN AQUACULTURE

Systems of environmental regulation of aquaculture based on development permits and discharge permits are conceptually inappropriate. Aquaculture development is of a start/stop or, more correctly, a fast/slow nature. Specific technological developments and the introduction of new species to aquaculture frequently result in rapid expansion of the sector in a very short space of time. Furthermore, because of its dependence on specific water resources, aquaculture development is often concentrated geographically.

Development permitting systems, even if based on EIA, have no mechanism to take account of the cumulative effects of rapid, geographically concentrated development, and can only act to constrain development retrospectively. The recognition within institutions that overdevelopment may occur, and the formulation and implementation of measures to curb that development, typically takes a number of years. By that time it is either too late, or too difficult administratively to take effective control for environmental purposes.

Discharge permits on their own are often viewed as a licence to pollute by the permit holder, and offer no incentive for reducing total pollution loads. The 'polluter pays' principle will go some way to redressing this situation, but its implementation is likely to be a long-term, incremental process.

Input management, along the lines of the recent Danish legislation described above, can bring about immediate benefits – for the fish farmer as well as with regard to environmental protection. It is, therefore, a more environmentally sound approach compared with fiscal measures based on pollution loads.

The achievement of **sustainability** at a national level and the implementation of appropriate legislation will inevitably take an incremental approach. Intrinsic to this process is that continuous improvements in environmental performance should be seen. This needs to be true for individual businesses as well as at a national level: indeed it may well be a prerequisite. Existing regulations need to be progressively upgraded to improve the quality of management (feeding strategies, husbandry), technology (feed formulation, wastewater treatment) and thus wastewater discharge quality. Provided this is backed by effective controls on development, these measures will ensure improved environmental health.

These changes need to be sensitive to the economic climate of the sector. Unnecessarily severe regulation would damage important industries and the credibility of environmental regulation and, ultimately, the whole concept of sustainable development.

A strategic regional approach, using SEA, can ensure that appropriate levels of development are achieved and sustained without long-term harm to the environment. In the Thai shrimp farming industry, for example, a development-led approach has created considerable environmental problems. Productivity of newly developed areas has been sustained for only a few years before being severely curtailed by environmental degradation (Chapter 2). A strategic approach would have recognized the carrying capacity of the coastal environment and would have limited development to an appropriate level, set with regard to total output and the density of development. It would also have recognized the contribution of best environmental practice and so implemented regulations concerning input management and waste disposal. This could have ensured the long-term sustainability of the industry, and the continued success of businesses that have instead been forced to cease operation.

SEA can provide an umbrella framework for many of the existing but relatively uncoordinated measures that can be seen in place around the world today. **Development permits** (e.g. British Columbia) would specify production limits or, more appropriately, limits to pollution loads. These would be based on **input management** measures (e.g. Denmark), to ensure that the discharge limits would not be exceeded due to poor management practices. The ecological **carrying capacity** (e.g. Denmark, Norway) of distinct ecological units (lakes, river catchments, coastal cells) would be determined, and **decision support systems** based on relevant environmental models might be used in development control. **Discharge permits** would be allocated within this framework, to ensure that the carrying capacity is not exceeded. Permits might also specify the adoption of **best environmental practice** and simple **environmental management systems** (elements of BS7750, British Standards Institute) to ensure continuing improvements in environmental performance, to coincide with the incremental strengthening of environmental regulation at a national or regional level.

The regional assessment of carrying capacity based on the ecological characteristics of local resources could ensure more appropriate development patterns. In more sensitive environments with a small assimilative capacity, consent conditions could be more stringent, and therefore more expensive to achieve or more limiting on production. This could direct development to less sensitive areas where development of aquaculture is more suited to the prevailing environmental characteristics. This is a transparent system. There are often to be found regional differences in environmental regulation, many of them without logical reasons, and this creates enmity, economic imbalance and a disrespect for the regulatory measures. Measures must be credible if they are to be successful, and an important component of SEA is its transparency and the involvement of industry and community.

SEA, undertaken with industry and community participation, should result in a fair and rational assessment of resources and regulation of development (Therivel *et al.*, 1992). For this reason it is in the interests of the aquaculture industry to participate fully in this process at an institutional level (e.g. local producer organizations, cooperatives, national associations). It could be suggested that it is in the long-term interests of the industry actively to encourage SEA of aquaculture. The potential consequences of not doing so include:

- failure of the industry through environmental problems;
- dissipation of economic benefit through excessive investment and overproduction (e.g. world salmon and shrimp farming);
- implementation of harsh environmental legislation (e.g. Denmark). Whilst the Danish system points the way forward in principle, many have commented that the specific limits imposed on fish farms are too severe (FES, 1992).

It should be noted that examples of each of these scenarios have already emerged. There is no doubt that they will be repeated globally if a strategic approach to aquaculture development is not taken by government and by the industry.

SUMMARY

- Regulation is necessary to control the impact on the environment of aquaculture development.

- Approaches to the regulation of aquaculture for environmental protection include:
  - development permits, specifying size and/or location of the fish farm;
  - environmental quality standards (achieved through nutrient loading limits and other conditions specified in consents to discharge);
  - waste minimization regulations, relating to feeding management;
  - discretionary requirement for EIA.
- Strategic approaches to aquaculture development in the context of environmental protection have been adopted in Denmark and Norway, for example. In these cases the level of permitted development is related to the carrying capacity of the environment for aquaculture.

References

Alanärä, A. (1992a) The effect of time-restricted demand feeding on feeding activity, growth and feed conversion in rainbow trout (*Oncorhynchus mykiss*). *Aquaculture* **108**, 357–368

Alanärä, A. (1992b) Demand feeding as a self regulating feeding system for rainbow trout (*Oncorhynchus mykiss*) in net pens. *Aquaculture* **108**, 347–356

Alanärä, A. (1994) The effect of temperature, dietary energy content and reward level on the demand feeding activity of rainbow trout (*Oncorhynchus mykiss*). *Aquaculture* **126**, 349–359

Alderman, D.J. and Michel, C. (1992) Chemotherapy in aquaculture today, in *Chemotherapy in Aquaculture: from theory to reality*, (eds C. Michel and D.J. Alderman), Office International des Epizooites, Paris

Aldridge, C. (1994) Making the most of 'green' credentials. *Fish Farmer* **2** (March/April), 8

Alsted, N. (1991) Studies on the reduction of discharges from fish farms by modification of the diet, in *Nutritional Strategies and Aquaculture Waste* (eds C.H. Cowey and C.Y. Cho), Proceedings of the First International Symposium on Nutritional Strategies in Management of Aquaculture Waste, University of Guelph, Ontario, Canada

Anon. (1994) A guide to the use of cleaner fish. *Fish Farmer* **17** (6), 7–8

Arbiv, R. and van Rijn, J. (1995) Performance of a treatment system for inorganic nitrogen removal in intensive aquaculture systems. *Aquacultural Engineering* **14**, 189–203

Armstrong J., Armstrong, W. and Beckett, P.M. (1992) *Phragmites australis*: Venturi- and humidity-induced pressure flows enhance rhizome aeration and rhizosphere oxidation. *New Phytology* **120**, 197–207

Bardach J.E., Ryther J.H. and McLarney, W.O. (1972) *Aquaculture: The Farming and Husbandry of Freshwater and Marine Organisms*, John Wiley and Sons, New York, 868 pp

Bergheim, A. and Asgard, T. (1996) Waste production from aquaculture, in *Aquaculture and Water Resource Management* (eds D.J. Baird, M.C.M. Beveridge and L.A. Kelly), Blackwell Science, Oxford.

Bergheim, A., Hustveit, H., Kittelsen, A. and Selmer-Olsen, A.R. (1984) Estimated pollution loadings from Norwegian fish farms II. Investigations 1980–1981. *Aquaculture* **36**, 157–168

Bergheim A., Aabel J.B. and Seymour, E.A. (1991) Past and present approaches to aquaculture waste management in Norwegian net pen culture operations, in *Nutritional Strategies and Aquaculture Waste* (eds C.B. Cowey and C.Y. Cho), Proceedings of the First International Symposium on Nutritional Strategies in Management of Aquaculture Waste, University of Guelph, Ontario, Canada

Bergheim, A., Sanni, S.,Indrevik, G. and Holland, P. (1993) Sludge removal from salmonid tank effluent using rotating microsieves. *Aquacultural Engineering* **12**, 97–109

Beveridge, M.C.M. (1987) *Cage Aquaculture*, Fishing News Books, Oxford, 351 pp

Biddlestone, A.J., Gray, K.R. and Thurairajan, K. (1991) A botanical approach to the treatment of wastewaters. *Journal of Biotechnology* **17**, 209–220

Boyd, C.E. (1995) *Bottom Soils, Sediment and Pond Aquaculture*, Chapman & Hall, New York, 341 pp

Boyd, C.E., Hollerman, W.D., Plumb, J.D. and Saeed, M. (1984) Effect of treatment with commercial bacterial suspension on water quality in channel catfish ponds. *Progressive Fish Culturist* **46**(1), 37–40

Briggs, M.R.P. (1994) Status, problems and solutions for a sustainable shrimp culture industry with special reference to Thailand, in *Development Strategies for Sustainable Shrimp Farming*, Report to ODA, Research Project R4751, 1–41

Briggs, M.R.P and Funge-Smith, S. (1994) A nutrient budget of some intensive marine shrimp ponds in Thailand, in *Development of Strategies for Sustainable Shrimp Farming*, Report to ODA, Research Project R4751, Stirling University, UK

Brown, J.R., Gowen, R.J. and McLusky, D.S. (1987) The effect of salmon farming on the benthos of a Scottish sea loch. *Journal of Experimental Marine Biology and Ecology* **109**, 39–51

Bruce, A.M. and Hawkes, H.A. (1983) Biological filters, in *Ecological Aspects of Used-Water Treatment* (eds C.R. Curds and H.A. Hawkes), Academic Press, London

Buschmann, A.H., Troell, M., Kautsky, N. and Kautsky, L. (1996) Integrated tank cultivation of salmonids and *Gracilaria chilensis* (Gracilariales, Rhodopyta). *Hydrobiologia* **326/327**, 75–82

Butz, I. (1992) Regulating waters from aquaculture in Austria, in *Workshop on Fish Farm Effluents and their Control in EC Countries, Hamburg, November 1992* (eds H. Rosenthal and V. Hilge), p.26.

Caine, G.D. and Castledine, J. (1991) Environmental regulations framed to minimise waste in aquaculture effluents in British Columbia, in *Nutritional Strategies and Aquaculture Waste* (eds C.B. Cowey and C.Y. Cho), Proceedings of the First International Symposium on Nutritional Strategies in Managment of Aquaculture Waste, University of Guelph, Ontario, Canada

Camargo, J.A. (1992) Temporal and spatial variations in dominance, diversity and biotic indices along a limestone stream receiving a trout farm effluent. *Water, Air and Soil Pollution* **63**, 343–359

Canfield, D.E. Jr and Bachmann, R.W. (1981) Prediction of total phosphorus concentrations, chlorophyll and Secchi depths in natural and artificial lakes. *Canadian Journal of Fisheries and Aquatic Sciences* **38**, 414–423.

Carlsson, S.-A. (1991) Fish feed and environmental impact: a review of the Swedish development during the last decade, in *Nutritional Strategies and Aquaculture Waste* (eds C.H. Cowey and C.Y. Cho), Proceedings of the First International Symposium on Nutritional Strategies in Management of Aquaculture Waste, University of Guelph, Ontario, Canada

Carr, O.J. and Goulder, R. (1990a) Fish farm effluents in rivers. I. Effects of bacterial populations and alkaline phosphatase activity. *Water Research* **24**(5), 631–638

Carr, O.J. and Goulder, R. (1990b) Fish farm effluents in rivers. II. Effects of inorganic nutrients, algae and the macrophyte *Ranunculus penicillatus*. *Water Research* **24**(5), 639–647

Chapra, S.C. (1975) Comment on 'An empirical method of estimating the retention of phosphorus in lakes' by W.B. Kirchner and P.J. Dillon. *Water Resource Research* **11**, 1033–1034.

Chen, S., Timmons, M.B., Bisogni, J.J. Jr and Aneshansley, D.J. (1993a) Protein and its removal by foam fractionation. *Progressive Fish Culturist* **55**, 76–82

Chen, S., Timmons, M.B., Bisogni, J.J. Jr and Aneshansley, D.J. (1993b) Suspended solids removal by foam fractionation. *Progressive Fish Culturist* **55**, 69–78

Christensen, V. and Pauly, D. (1992) ECOPATH II – a software for balancing steady-state ecosystem models and calculating network characteristics. *Ecological Modeling* **61**, 169–185

Chua, T.E., Paw, J.N. and Guarin, F.Y. (1989) The environmental impact of aquaculture and the effects of pollution on coastal aquaculture development in southeast Asia. *Marine Pollution Bulletin* **20**, 335–343

Clark, E.R., Harman, J.P. and Forster, J.R.M. (1985) Production of metabolic and waste products by intensively farmed rainbow trout *Salmo gairdneri* Richardson. *Journal of Fish Biology* **27**, 381–393

Colt, J.E. and Tchobanoglous, G. (1981) Design of aeration systems for aquaculture, in *Proceedings of the Bioengineering Symposium for Fish Culture*, (eds L.J. Allen and E.C. Kinney), Fish Culture Section Publication 1, American Fisheries Society, Bethesda, Maryland, pp. 138–148

Cooper, P.F. and Findlater, B.C. (eds) (1990) *Constructed Wetlands in Water Pollution Control*, International Association on Water Pollution Research and Control, Pergamon Press, Oxford

Coyne, R., Hiney, M., O'Connor, B. *et al.* (1994) Concentration and persistence of oxytetracycline in sediments under a marine salmon farm. *Aquaculture* **123**, 31–42

Cripps, S.J. (1994) Minimising outputs – treatment. *Journal of Applied Ichthyology – Zeitschrift für Angewandte Ichthyologie* **10**, 284–294.

Cripps, S.J. and Kelly, L. (1996) Reductions in wastes from aquaculture, in *Aquaculture and Water Resource Management* (eds D. Baird, M.C.M. Beveridge, L. Kelly and J. Muir), Blackwell Science, Oxford, pp. 166–201

Csavas, I. (1990) Shrimp aquaculture developments in Asia, in *Technical and Economic Aspects of Shrimp Farming* (eds M.B. New, H. de Saran and T. Singh), Proceedings of the Aquatec '90 Conference, Kuala Lumpur, Malaysia, 1990

Dahlbäck, B. and Gunnarsson, L.Å.H. (1981) Sedimentation and sulfate reduction under a mussel culture. *Marine Biology* **63**, 269–275

Dalsgaard, J.P.T., Lightfoot, C. and Christensen, V. (1995) Towards quantification of ecological sustainability in farming systems analysis. *Ecological Engineering* **4**, 181–189

Dierberg, F.E. and Kiattisimkul, W. (1996) Issues, impacts and implications of shrimp culture in Thailand. *Environmental Management* **20**, 649–666

EIFAC (1982) *Report of the EIFAC Workshop on Fish-Farming Effluents, Silkeborg, Denmark*, (ed. J.S. Alabaster), EIFAC Technical Paper 41, FAO, Rome, pp. 113–166

EIFAC (1988) National contributions on suspended solids from land-based fish farms (ed. M. Pursiainan). Papers presented at the first session of the EIFAC Working Party on Fish Farm Effluents, The Hague, Netherlands, May/June 1987. Riista–Ja Kalatalouden Tutkimuslaitos, Kalantutkimusosasto, Monistettuja, Julkkaisuja, Helsinki

Etnier C. and Guterstam, B. (eds) (1991) *Ecological Engineering for Wastewater Treatment*, Proceedings of a conference at Stensund Folk College, Sweden, March 24–28, 1991. Bokskogen, Sweden (Publ.)

European Commission (1995) *Aquaculture and the Environment in the European Community*, Office for Official Publications of the European Communities, Brussels

Falconer, R.A. and Hartnett, M. (1993) Mathematical modelling of flow, pesticide and nutrient transport for fish farm planning and management. *Ocean and Coastal Management* **19**, 37–57

FES (1992) *Fish Farm Effluents and their Control in EC Countries*, FES position paper, Fédération Européenne de la Salmoniculture, London.

Folke, C. and Kautsky, N. (1989) The role of ecosystems for a sustainable development of aquaculture. *Ambio* **18**(4), 234–243

Folke, C. and Kautsky, N. (1992) Aquaculture with its environment: prospects for sustainability. *Ocean and Coastal Management* **17**, 5–24

Funge-Smith, S.J. and Briggs, M.R.P. (1994) Growth and ammonia removal by *Gracillaria* spp. in brackishwater and intensive shrimp farm influents and effluents, in *Development Strategies for Sustainable Shrimp Farming*, Report to ODA, Research Project R4751, pp. 1–41

GESAMP (1991) *Reducing Environmental Impacts of Coastal Aquaculture*, Reports and Studies 47, GESAMP (Joint Group of Experts on the Scientific Aspects of Marine Pollution), FAO, Rome, 35 pp

GESAMP (1997) *Towards safe and effective use of chemicals in coastal aquaculture*, Reports and studies 65, GESAMP (Joint Group of Experts on the Scientific Aspects of Marine Environmental Protection), FAO, Rome, 40 pp

Gowen, R.J., Bradbury, N.B. and Brown, J.R. (1989) The use of simple models in assessing two of the interactions between fish farming and the marine environment, in *Aquaculture – a Biotechnology in Progress* (eds N. DePauw, E. Jaspers, H. Akefors and N. Wilkins), European Aquaculture Society, Bredene, Belgium.

Gray, N.F. (1989) *Biology of Wastewater Treatment*, Oxford University Press, Oxford.

Gumbricht, T. (1993) Nutrient removal processes in freshwater submersed macrophyte systems. *Ecological Engineering* **2**, 1–30

Hankin, R. (1992) Veterinary medicines regulations in the EEC, in *Chemotherapy in Aquaculture: from theory to reality*, (eds C. Michel and D.J. Alderman), Office International des Epizooites, Paris

Hargrove, L.L., Westerman, P.W. and Losordo, T.M. (1995) Nitrification in three-stage and single-stage floating bead biofilters in a laboratory-scale recirculating aquaculture system. *Aquacultural Engineering* **15**, 67–80

Henderson, J.P. and Bromage, N.R. (1988) Optimising the removal of suspended solids from aquacultural effluents in settlement lakes. *Aquacultural Engineering* **7**, 167–181

Hepher, B. (1988) *Nutrition of Pond Fishes*, Cambridge University Press, Cambridge.

Hess, W.J. (1981) Performance ratings for submerged nitrification biofilters: development of a design calculation procedure, in *Bio-Engineering Symposium for Fish Culture*, Fish Culture Section of the American Fisheries Society Publication 1, Bethesda, Maryland, USA, pp. 63–70

Hiley, P. (1990) Wetlands treatment revival in Yorkshire, in *Constructed Wetlands in Water Pollution Control* (eds P.F. Cooper and B.C. Findlater), International Association on Water Pollution Research and Control, Pergamon Press, Oxford

Hopkins, J.S., Brady, C.L., Hamilton, R.D. and Heffernan, J.A. (1995) The effect of low-rate sand filtration and modified feed management on effluent quality, pond water quality and production of intensive shrimp ponds. *Estuaries* **18**, 116–123

Howarth, W. (1990) *The Law of Aquaculture*, Fishing News Books, Oxford

HRPB (1987) *Policy for Controlling Pollution from Fish Farms*, Highland River Purification Board, Dingwall

Ibrekk, H.O., Kryvi, H. and Elvestad, S. (1993) Nationwide assessment of the suitability of the Norwegian coastal zone and rivers for aquaculture (LENKA). *Coastal Management* **21**, 53–73

Jackson, P. and Ashton, D. (1993) *Implementing Quality Through BS5750 (ISO 9000)*, Kogan Page Ltd, 225 pp

Jensen, J.B. (1991) Environmental regulation of fresh water fish farms in Denmark, in *Nutritional Strategies and Aquaculture Waste* (eds C.B. Cowey and C.Y. Cho), Proceedings of the First International Symposium on Nutritional Strategies in Management of Aquaculture Waste, University of Guelph, Ontario, Canada

Jobling, M. (1993) Nutrition, diet formulation and feeding practices, in *Salmon Aquaculture* (eds K. Heen, R.L. Monahan and F. Utter), Fishing News Books, Oxford

Johnsen, F. and Wandsvik, A. (1991) The impact of high energy diets on pollution control in the fish farming industry, in *Nutritional Strategies and Aquaculture Waste* (eds C.H. Cowey and C.Y. Cho), Proceedings of the First International Symposium on Nutritional Strategies in Management of Aquaculture Waste, University of Guelph, Ontario, Canada

Jones, J.R. and Bachmann, R.W. (1976) Prediction of phosphorus and chlorophyll levels in lakes. *Journal of the Water Pollution Control Federation* **48**, 2176–2182

Jungersen, G. (1991) Resource-saving and ecological aspects of using wastewater from eel breeding as a source of nutrients, water and carbon dioxide for plant production, in *Ecological Engineering for Waste Water Treatment*, Proceedings of a conference at Stensund Folk College, Sweden, March 24–28, 1991. Bokskogen, Sweden (Publ.)

Karunasagar, I., Pai, R., Malathi, G.R. and Karunasagar, I. (1994) Mass mortality of *Penaeus monodon* larvae due to antibiotic-resistant *Vibrio harveyi* infection. *Aquaculture* **128**, 203–209

Kautsky, N. and Folke, C. (1991) Integrating open system aquaculture. *Ecological Engineering for Waste Water Treatment*, Proceedings of a conference at Stensund Folk College, Sweden, March 24-28, 1991. Bokskogen, Sweden (Publ.)

Kerry, K., Hiney, M., Coyne, R. *et al.* (1994) Frequency and distribution of resistance to oxytetracycline in micro-organisms isolated from marine fish farm sediments following therapeutic use of oxytetracycline. *Aquaculture* **123**, 43–54

Kiaerskou, J. (1991) Production and economics of 'low pollution diets' for the aquaculture industry, in *Nutritional Strategies and Aquaculture Waste* (eds C.H. Cowey and C.Y. Cho), Proceedings of the First International Symposium on Nutritional Strategies in Management of Aquaculture Waste, University of Guelph, Ontario, Canada

Kitoh, S., Shiomi, X. and Uheda, E. (1993) The growth and nitrogen fixation of *Azolla filiculoides* Lam. in polluted water. *Aquatic Botany* **46**, 129–139

Kupka-Hansen, P., Lunestad, B.T. and Samuelsen, O.B. (1992) Ecological effects of antibiotics and chemotherapeutants from fish farming, in *Chemotherapy in Aquaculture: from theory to reality*, (eds C. Michel and D.J. Alderman), Office International des Epizooites, Paris

Lakshman, G. (1979) An ecosystem approach to the treatment of waste waters. *Journal of Environmental Quality* **8**(3), 353–361

Larsen, D.P. and Mercier, H.T. (1976) Phosphorus retention capacity of lakes. *Journal of the Fisheries Research Board of Canada* **33**, 1742–1750

Lawson, T.B (1995) *Fundamentals of Aquaculture Engineering*, Chapman & Hall, 351 pp

Lewis, W.M. (1980) *Developments in Water Treatment* (2 vols), Applied Science Publishers, London.

Ley, C. (1993) Ardvars answer – site rotation. *Fish Farmer* **5** (September/October 1993).

Lightfoot, C., Bimbao, M.A., Dalsgaard, J.P.T. and Pullin, R.S.V. (1993) Aquaculture and sustainability through integrated resources management. *Outlook on Agriculture* **22**(3), 143–150

Liltved, H., Hektoen, H. and Efraimsen, H. (1995) Inactivation of bacterial and viral fish pathogens by ozonation or UV irradiation in water of different salinity. *Aquacultural Engineering* **14**, 107–122

Little, D. and Muir, J. (1987) *Integrated Warm Water Aquaculture*, Institute of Aquaculture Publications, University of Stirling.

Lunestad, B.T. (1992) Fate and effects of antibacterial agents in aquatic environments, in *Chemotherapy in Aquaculture: from theory to reality*, (eds C. Michel and D.J. Alderman), Office International des Epizooites, Paris

Magurran, A.E. (1988) *Ecological Diversity and its Measurement*, Chapman & Hall, London.

Makinen, T. (1983) Means to decrease loadings from fish farms, to separate sludge and to treat sludge liquor. Seminar paper presented at Vesitutkimuspaivat, October 1983, University of Kuopio. (In Finnish.)

Makinen. T., Lindgren, S. and Eskelinen, P. (1988) Sieving as an effluent treatment method for aquaculture. *Aquacultural Engineering* **7**, 367–377

Marking, L.L., Leith, D. and Davis, J. (1990) Development of a carbon filter system for removing malachite green from hatchery effluents. *Progressive Fish Culturist* **52**, 92–99

Mattson, J. and Linden, O. (1983) Benthic macrofauna succession under mussels *Mytilus edulis* L. (Bivalvia), cultured on hanging long-lines. *Sarsia* **68**, 97–102

May, E., Butler, J.E., Ford, M.G. *et al.* (1990) Chemical and microbiological processes in gravel-bed hydroponic (GBH) systems for sewage treatment. In *Constructed Wetlands in Water Pollution Control* (eds P.F. Cooper and B.C. Findlater), International Association on Water Pollution Research and Control, Pergamon Press, Oxford

McMurty, M.R., Sanders D.C., Patterson R.P. and Nash, A. (1993a) Yield of tomato irrigated with recirculating aquaculture water. *Journal of Production Agriculture* **6**(3), 428–432

McMurty, M.R., Sanders D.C., Patterson R.P. and Nash, A. (1993b) Mineral nutrient concentration and uptake by tomato irrigated with recirculating aquaculture water as influenced by quantity of fish waste products supplied. *Journal of Plant Nutrition* **16**(3), 407–419

Merican, Z.O. and Phillips, M.J. (1985) Solid waste production from rainbow trout, *Salmo gairdneri* Richardson, cage culture. *Aquaculture and Fisheries Management* **1**, 55–69

Michel, C. and Alderman, D.J. (eds) (1992) *Chemotherapy in Aquaculture: theory to reality*, Office Internationale des Epizooites, Paris

Mitsch, W.J. and Jorgensen, S.E. (eds) (1989) *Ecological Engineering – an introduction to ecotechnology*, John Wiley and Sons, New York

Moriarty, D.J.W and Body, A.G.C. (1995) Modifying microbial ecology in ponds: a key to sustainable aquaculture. *Asia Aquaculture, Singapore*, September 1995

Moss, B. (1980) *Ecology of Fresh Waters*, Blackwell Scientific, Oxford

Nash, C.E. (1995) *Aquaculture Sector Planning and Management*, Fishing News Books, Oxford

NCC (1990) *Fish Farming and the Scottish Freshwater Environment*, Report for the Nature Conservancy Council by Institute of Aquaculture, Stirling, Institute of Freshwater Ecology (NERC) and Institute of Terrestrial Ecology (NERC).

Ng, W.J, Sim, T.S. and Ong, K.K. (1990) The effect of *Elodea densa* on aquaculture water quality. *Aquaculture* **84**, 267–276

Nijhof, M. (1995) Bacterial stratification and hydraulic loading effects in a plug-flow model for nitrifying trickling filters applied in recirculating fish culture systems. *Aquaculture* **134**, 49–64

Odum, E.P. (1993) *Ecology: Our Endangered Life-Support Systems,* Sinauer Associates, Sunderland, Massachusetts, 301 pp

O'Laoire, D. and Welford, R. (1994) Environmental management at the regional level: the case of the Avoca–Avonmore Catchment Conversion Project and IDAS Trout Ltd, in *Cases in Environmental Management and Business Strategy* (ed. R. Welford), Pitman Publishing, London

Ormaza-Gonzalez, F.I. and Vilalba-Flor, P. (1994) The measurement of nitrite, nitrate and phosphate with test kits and standard procedures: a comparison. *Water Research* **28**(10) 2223–2228

Ostrofsky, M.L. (1978) Modification of phosphorus retention models for use with lakes with low areal loading. *Journal of the Fisheries Research Board of Canada* **35**, 1532–1536

Pauli, G. (1994) ZERI – Outline of the feasibility studies and partners 1 Beer/ Algae/Fish Farming 2 Forestry/Green Mass. Unpublished report, United Nations University, Tokyo

Pedersen, Chr.L. (1987) Energy budgets for juvenile rainbow trout at various oxygen concentrations. *Aquaculture* **62**, 289–298

Penczak, T., Galicka, W., Molinski, E. *et al.* (1982) The enrichment of a mesotrophic lake by carbon, phosphorous and nitrogen from the cage aquaculture of rainbow trout, *Salmo gairdneri*. *Journal of Applied Ecology* **19**, 371–393

Persson, G. (1991) Eutrophication resulting from salmonid fish culture in fresh and salt waters: Scandinavian experiences, in *Nutritional Strategies and Aquaculture Waste,* (eds C.H. Cowey and C.Y. Cho), Proceedings of the First International Symposium on Nutritional Strategies in Management of Aquaculture Waste, University of Guelph, Ontario, Canada

Phillips, M.J. (1985) The environmental impact of cage culture on Scottish freshwater lochs. Report to the Highlands and Islands Development Board, Institute of Aquaculture, University of Stirling

Pillay, T.V.R. (1990) *Aquaculture: Principles and Practices,* Fishing News Books, Oxford, 575 pp

Pillay, T.V.R. (1992) *Aquaculture and the Environment,* Fishing News Books, Oxford, 189 pp

Pond, W.G. and Mumpton, F.A. (eds) (1984) *Zeo-Agriculture: the use of natural zeolites in agriculture and aquaculture,* Westview Press, Boulder, Colorado

Poxton, M.G. (1991) Water quality fluctuations and monitoring in intensive fish culture, in *Aquaculture and the Environment* (eds N. de Pauw and J. Joyce), European Aquaculture Society Special Publication No. 6, Gent, Belgium

Poxton, M.G. and Allouse, S.B. (1987) Cyclical fluctuations in ammonia and nitrate-nitrogen resulting from the feeding of turbot, *Scopthalmus maximus* (L), in recirculating systems. *Aquacultural Engineering* **6**(4), 301–322

Primavera, J.H. (1991) Intensive prawn farming in the Philippines: ecological, social and economic implications. *Ambio* **20**(1), 28–33

Rakocy, J.E. and Allison, R. (1981) Evaluation of a closed recirculation system for the culture of tilapia and aquatic macrophytes. *Bio-Engineering Symposium for Fish Culture,* Fish Culture Section of the American Fisheries Society Publication 1, Bethesda, Maryland, USA, pp. 296–307

Redding, T., Todd, S. and Midlen, A. (1997) The treatment of aquaculture wastewaters – a botanical approach. *Journal of Environmental Management* **50**, 283–299

Reddy, K.R and De Busk, W.F. (1985) Nutrient removal potential of selected aquatic macrophytes. *Journal of Environmental Quality* **14**(4), 459–463

Reddy K.R., D'Angelo, E.M and De Busk, T.A (1989) Oxygen transport through aquatic macrophytes: the role in wastewater treatment. *Journal of Environmental Quality* **19**, 261–267

Richards, R.H. (1992) Constraints on drug prescription – a veterinary viewpoint, in *Chemotherapy in Aquaculture: from theory to reality* (eds C. Michel and D.J. Alderman), Office International des Epizooites, Paris

Ricklefs, R.E. (1979) *Ecology*, 2nd edn, Thomas Nelson and Sons, Sunbury-on-Thames

Rodhouse, P.G., Roden, C.M., Hensey, M.P. and Ryan, T.H. (1985) Production of mussels *Mytilus edulis* in suspended culture and estimates of carbon and nitrogen flow: Killary Harbour, Ireland. *Journal of the Marine Biological Association of the UK* **65**, 55–68

Røed, H. (1991) Demands for environmental testing of antibacterials to be used in Norwegian fish farming and environmental solutions. *Aquaculture and the Environment* **16**, 265–276

Samuelsen, O.B. (1992) The fate of antibiotics/chemotherapeutics in marine aquaculture sediments, in *Chemotherapy in Aquaculture: from theory to reality* (eds C. Michel and D.J. Alderman), Office International des Epizooties, Paris

Schwartz, M.F. and Boyd, C.E. (1995) Constructed wetlands for treatment of channel catfish pond effluents. *Progressive Fish Culturist* **57**(4), 255–266

Scott, K.R. and Allard, L. (1983) High-flowrate water recirculation system incorporating a hydroclone prefilter for rearing fish. *Progressive Fish Culturist* **45**(3), 148–153

Sedgewick, S.D. (1990) *Trout Farming Handbook*, Fishing News Books, Oxford, 164 pp

Shepherd, C.J. and Bromage, N.R. (1988) *Intensive Fish Farming*, BSP Professional Books, Blackwell Science, Oxford, 404 pp

Shpigel, M. and Neori, A. (1996) The integrated culture of seaweed, abalone, fish and clams in modular intensive land-based systems. I. Proportions of size and expected revenues. *Aquacultural Engineering* **15**(5), 313–326

Shpigel, M., Neori, A., Popper, D.A. and Gordin, H. (1993) A proposed model for environmentally clean land-based culture of fish, bivalves and seaweeds. *Aquaculture* **117**, 115–128

Silvert, W. (1992) Assessing environmental impact of finfish aquaculture in marine waters. *Aquaculture* **107**, 67–79

Smith, P., Donlon, J., Coyne, R. and Cazabon, D.J. (1994) Fate of oxytetracycline in a freshwater fish farm: influence of effluent treatment systems. *Aquaculture* **120**, 319–325

Spanggaard, B., Jorgensen, F., Gram, L. and Huss, H.H. (1993) Antibiotic resistance in bacteria isolated from three freshwater fish farms and an unpolluted stream in Denmark. *Aquaculture* **115**, 195–207

Speece, R.E. (1981) Management of dissolved oxygen and nitrogen in fish hatchery waters, in *Bio-Engineering Symposium for Fish Culture*, Fish Culture Section of the American Fisheries Society Publication 1, Bethesda, Maryland, USA, pp. 53–62

Tebbut, T.H.Y. (1992) *Principles of Water Quality Control*, 4th edn, Pergamon Press, Oxford, 251 pp

Therivel, R., Wilson, E. and Thompson, S. (1992) *Strategic Environmental Assessment*, Earthscan Publications Ltd, London

Tivy, J. and O'Hare, G. (1981) *Human Impact on the Ecosystem*, Oliver and Boyd, Edinburgh and New York, 243 pp

Tvinnereim, K. and Skybakmoen, S. (1989) Water exchange and self-cleaning in fish rearing tanks, in *Aquaculture – a Biotechnology in Progress* (eds N. DePauw, E. Jaspers, H. Ackefors and N. Wilkins), European Aquaculture Society, Bredene, Belgium

Vollenweider, R.A. (1975) Input–output models with special reference to the phosphorus loading concept in limnology. *Schweicz. Z. Hydrologi* **37**, 455–472

Wagner, E.J. (1993) Evaluation of a new baffle design for solid waste removal from hatchery raceways. *Progressive Fish Culturist* **55**, 43–47

Waller, J., Allen, D. and Burns, A. (1993) *The Quality Management Manual*, Kogan Page, London, 206 pp

Warrer-Hansen, I. (1982) Evaluation of matter discharged from trout farming in Denmark, in *Report of the EIFAC Workshop on Fish-Farming Effluents, Silkeborg, Denmark* (ed. J.S. Alabaster), EIFAC Technical Paper 41, pp. 57–63

Warrer-Hansen, I. (1982) Methods of treatment of waste water from trout farming, in *Report of the EIFAC Workshop on Fish-Farming Effluents, Silkeborg, Denmark* (ed. J.S. Alabaster), EIFAC Technical Paper 41, pp. 113–166

Warrer-Hansen, I. (1989) Production and control of wastes in freshwater fish farming. *Aquaculture Ireland* **40**, 19–22

Watten, B.J. (1990) Design of packed columns for commercial oxygen addition and dissolved nitrogen removal based on effluent criteria. *Aquacultural Engineering* **9**, 305–328

Weeks, N.C., Timmons, M.B. and Chen, S. (1992) Feasibility of using foam fractionation for the removal of dissolved and suspended solids from fish culture units. *Aquacultural Engineering* **11**, 251–265

Welford, R. (1994) *Cases in Environmental Management and Business Strategy*, Pitman Publishing, London

Welford, R. and Gouldson, A. (1993) *Environmental Management and Business Strategy*, Pitman Publishing, London, 219 pp

Welsch, W. (1992) A new concept for semi-intensive cage-co-culture of salmonids and blue mussels, in *Workshop on Fish Farm Effluents and their Control in EC Countries* (eds H. Rosenthal and V. Hilge), November 1992, Hamburg

Wenhua L. (1995) Feasibility study on brewing–aquaculture–agriculture zero emission research initiatives (ZERI-BAG). Unpublished report submitted to the United Nations University (UNU)

Westerman, P.W., Losordo, T.M. and Wildhaber, M.W. (1996) Evaluation of various biofilters in an intensive recirculating fish production facility. *Transactions of the American Society of Agricultural Engineers* **39**(2), 723–727.

Westers H. (1991) Operational waste management in aquaculture effluents, in *Nutritional Strategies and Aquaculture Waste* (eds C.H. Cowey and C.Y. Cho), Proceedings of the First International Symposium on Nutritional Strategies in Management of Aquaculture Waste, University of Guelph, Ontario, Canada

Wheaton, F.W. (1977) *Aquacultural Engineering*, John Wiley and Sons, Chichester

Willoughby, L.G. and Roberts, R.J. (1992) Towards strategic use of fungicides against *Saprolegnia parasitica* in salmonid fish hatcheries. *Journal of Fish Diseases* **15**, 1–13

Wolverton, B.C., McDonald, R.C. and Duffer, W.R. (1983) Microorganisms and higher plants for wastewater treatment. *Journal of Environmental Quality* **12**, 236–243

World Bank (1991) *Environmental Assessment Sourcebook. Vol. I: Policies, procedures and cross-sectoral issues*, Technical Paper 139, World Bank, Washington DC

Index

Page numbers in *italic* refer to tables.